INFAILIBLE

INFAILIBLE

The Artificial Intelligence Ideology
Reshaping Consumer Behavior

CHRIS HOOD

CH DIGITAL VENTURES
Franklin, Tennessee

No part of this book my be used or reproduced in any manner whatsoever without written permission except in the case of brief quotations embodied in critical articles and reivews, or where otherwise is referenced through original source materials. For permission requests, write to the publisher, addressed "Attention: Permissions Coordinator," at the address below:

CH Digital Ventures
2000 Mallory Lane, Suite 290 #1344
Franklin, TN 37067
chrishood.com

Although every effort has been made to ensure the accuracy and reliability of the information presented in this book, neither the author nor the publisher are liable for any loss or damages, incidental or consequential, that may arise directly or indirectly from the use or application of this information. All names, characters, businesses, places, events, and incidents have been anonymized unless explicitly stated otherwise.

The strategies and advice contained in this book do not guarantee success and the results may vary from individual to individual. The reader is responsible for understanding and accepting the inherent risk that results may differ.

Furthermore, the publisher and author disclaim all liability to the maximum extent permitted by law, in the event that the information, commentary, analysis, opinions, advice, and/or recommendations in this book prove to be inaccurate, incomplete, or unreliable, or result in any investment or other losses. The reader is advised to exercise their judgment and verify information before making any decisions based on the content of this book.

Library of Congress Cataloging-in-Publication Data

Names: Hood, Chris, author.
Title: Infailible: The Artificial Intelligence Ideology Reshaping Consumer Behavior.
Description: First Edition. | Tennessee : CH Digital Ventures, 2025.
Identifiers: LCCN 2025900160
   ISBN 979-8-9885384-6-2 (HC) | 979-8-9885384-8-6 (PB)
Subjects: LCSH: Consumer Behavior | Artificial Intelligence | Business
LC record available at https://lccn.loc.gov/2025900

To Elliott,
may your college years
open a world of possibilities.

CONTENTS

Infailible (adj.) - *the false sense of perfection attributed to artificial systems, especially AI, which blinds humans to its inherent flaws, limitations, and inability to replicate the complexity of human intelligence.*

EXPLORATION OF HUMAN INTELLIGENCE VS ARTIFICIAL INTELLIGENCE

Chapter	Customer Segment	Human Intelligence	AI Explored	AI Category
1	Customer Strategy	Aspirational Intelligence	Predictive AI	Limited Memory AI
2	Customer Insights	Logical Intelligence	Analytical AI	Narrow AI
3	Customer Transformation	Behavioral Intelligence	Agentic AI	Artificial General Intelligence (AGI)
4	Customer Experience	Emotional Intelligence	Affective AI	Theory of Mind AI
5	Customer Journeys	Spatial Intelligence	Vision AI	Limited Memory AI
6	Customer Relations	Cultural Intelligence	Contextual AI	Theory of Mind AI
7	Customer Engagement	Interpersonal Intelligence	Conversational AI	Limited Memory AI
8	Customer Success	Intrapersonal Intelligence	Sentiment AI	Narrow AI
9	Customer Loyalty	Moral Intelligence	Ethical AI	Meta AI (Governance AI)
10	Customer Operations	Practical Intelligence	Applied AI	Narrow AI
11	Customer Alignment	Collaborative Intelligence	Swarm AI	Collective AI
12	Customer Management	Naturalistic Intelligence	Environmental AI	Narrow AI
13	Customer Growth	Existential Intelligence	Philosophical AI	Artificial Superintelligence (ASI)

Introduction

HYPE AND CIRCUMSTANCE

I should have written this book sooner.

Sitting at my desk surrounded by Post-it notes full of new ideas, I'm reflecting on a roundtable webinar I participated in about the convergence of Artificial Intelligence (AI) and Customer Experience (CX). In my head, around 99.9% of the group of 200 people that tuned in believe AI is the best tool ever for developing richer customer experiences. I'm confident I was the remaining .1% who considers the influence of AI on customers leads them to feel disconnected from your company.

During recent months, I have become increasingly surprised by how people interpret AI or, more specifically, understand and believe in AI. When I say "believe in AI," I don't mean simple trust in its accuracy or abilities. I'm referring to a deeper ideological belief in the transformative power of AI—a system of thought that elevates AI as a key force in redefining human potential, ethics, and progress, much like belief systems that ascribe infallibility or purpose to a higher power. This infallible belief exists despite the technical reality of what AI is actually capable of doing. The belief in what AI can achieve far exceeds its current technical capabilities, creating an approximate seven-year gap between perception and reality. There is too much *Star Trek*, not enough *Columbo*.

What's even more amusing is scrolling through countless articles praising the wonders of AI—only to realize they were written by AI. It's like a chatbot patting itself on the back for being "revolutionary." The irony practically writes itself. On a positive note, the hashtag #degenerativeAI has grown since the fall of 2024. Degenerative AI highlights the risks of systems trained on repetitive, low-quality, or biased data, leading to a gradual decline in creativity, accuracy, and meaningful insights over time. This trend highlights a skeptical shift in the public's perception of AI, recognizing that the relentless push to use it in a variety of industries and situations isn't without consequences. While people are just beginning to question the integrity and direction of some AI implementations, they're still captivated by the possibilities, unable or unwilling to step back and take a broader look at the ramifications.

Putting the AI hype aside, the parallels to customer relationships are hard to ignore. I'm increasingly shocked by the number of organizations that co-opt the label "customer first," but fail to practice that mindset. Instead of embracing how AI technology benefits their customers, they instead focus on its magical potential for the benefit of their teams. As I described in my 2023 book, *Customer Transformation*, to be obsessed with your customers means that technology always comes last.

I had my first encounter with AI in 1999. As most of my followers know, I began my career at a movie theater and in my spare time took a second job in technology building websites and software tools.

At the theater, I had the privilege of entering auditoriums before the lights dimmed to welcome guests. Sharing news about upcoming movies and highlighted marketing campaigns, I'd get them engaged with trivia questions and movie posters or T-shirts for correct answers.

During this time, most theaters in the United States had on-screen advertisements prior to the film, similar to what you see today. However, in the 1990s they were 35mm slide-based, with canisters of images that looped automatically with a distinctive sound as each slide dropped into the viewfinder before being projected on the screen.

While walking into an auditorium to welcome the audience one day, a couple stopped me to chat and ask questions. They were Friday evening regulars (a trend that has died in the digital age), and I had gotten to know them over several months. Carl pointed to the screen and asked, "Do they ever update these slides?" I laughed and explained that a few are replaced each month. [As a side note for web enthusiasts: the image carousels used on websites today get their name from the vintage Kodak 35mm slide carousel patented in 1965. This is the same year Joseph Weizenbaum created ELIZA, the world's first chatbot, which pioneered natural language processing by mimicking a Rogerian psychotherapist.]

Finding an answer for Carl's question was more complicated than a "yes" or "no." The basic formula for these on-screen advertisement packages included local business advertisements, theater concessions promotions, upcoming movie announcements, and trivia questions, with the answers separated by additional advertisements. An average carousel held about 100 slides, and each slide appeared on the screen for about 15 seconds before automatically advancing to the next. On average, the entire presentation was approximately 25 minutes long.

For those regular patrons, arriving at the theater 30 minutes before showtime meant you'd see the same slides. And if you visited multiple times a month, it was likely that you'd memorized the slides. I often heard people shouting out the answers to trivia questions in the theater—not because they knew the answer immediately, but rather because they had already seen those slides previously or even in the same evening. Other faults plagued the carousel system. It wasn't uncommon to find a slide flipped upside down or backward due to human error and manual processes for updating the presentation.

Reflecting on the conversation with Carl, I asked myself an important question: could I build a digital system that prevented typical errors and updated content automatically? Better yet, could I make the system intelligent so that no trivia question was asked more than once and adjusted based on who was in the audience? Could I also make it interactive?

After a few months of development, we premiered Digireel, the first-ever digital entertainment and advertising system, at Cinemapolis movie theater in Anaheim Hills, California (it was convenient I worked there). It leveraged digital projectors, an on-premise server, a connected database of advertisements, hundreds of trivia questions, and other forms of interactive entertainment and games. We introduced animations, sound, video, and an intelligent interactive mechanism to personalize the content based on the movie being shown in a specific theater, its MPA rating, time of day, and audience demographic. Honestly, this system was a bit ahead of its time. However, after I obtained a patent, the product was acquired by National CineMedia, Inc., and became the foundation for digital entertainment you see prior to movies today—albeit with more ads, less entertainment, and no intelligence. Over the years, as mobile technology, data tools, and AI grew in accessibility and capability, I've frequently thought about rebuilding Digireel. Despite the drastic decline in movie theater attendance leading to an unrealized investment, I imagine a powerful upgraded experience with today's AI. And that's the moral of this story: Despite the power and availability of AI, there is no new customer problem to solve or consumer demand for an upgraded advertising platform powered by AI at movie theaters. To do so would be a focus on technology, not the customer.

The Bandwagon Effect

This phenomenon, not limited to the political arenas where it first took root, now infiltrates corporate boardrooms with a potent force. Despite the marketing benefits of this cognitive bias, a crucial misstep is occurring in today's digital strategy, where AI has become a buzzword synonymous with innovation and forward-thinking. Executives, captivated by the widespread availability of AI, are driving its integration not for its genuine benefits but because it's trendy—or perhaps, more cynically, as a means to enable complacency.

Suddenly, I witnessed hundreds of people become "AI Experts" and watched sales pitches flood my inbox with "one-of-a-kind" AI tools that could solve all my problems. This gold rush toward AI solutions has left the market saturated with questionable products and diluted the genuine understanding and potential applications of AI technology. Everywhere you look, there are bold claims about AI transforming businesses overnight, yet few address the nuanced challenges or the strategic integration necessary for these tools to deliver on those promises. This hasty jump onto the AI bandwagon has fostered a landscape where hype often overshadows substance, leading to disillusionment and skepticism among consumers who are promised revolution but experience little real change.

PEOPLE AND ARTIFICIAL INTELLIGENCE

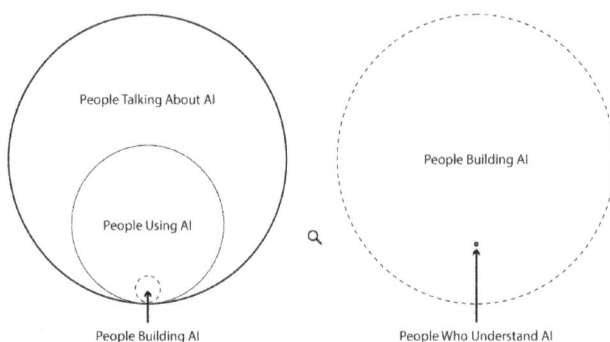

This gap between hype and reality is reflected in the above diagram. The largest circle on the left represents the overwhelming number of people talking about AI, driven by media buzz and speculative claims. Within that, a smaller circle shows those actively using AI—leveraging tools for tasks or integrating them into workflows. An even smaller circle depicts the people building AI: the researchers, developers, and engineers creating these systems. On the right, zooming in, a tiny dot within the builders that represents those who truly understand AI. It's worth noting that some individuals who genuinely understand AI may not be actively building, using, or discussing it for reasons explored in this book. These experts possess a profound understanding of AI's technical, ethical, and societal dimensions, enabling them to distinguish between realistic possibilities and inflated promises. The diagram highlights how widespread discussions about AI often outpace genuine comprehension, leading to misconceptions and misplaced expectations.

In December 2024, the rapid shift from buzzwords "generative AI" to "agentic AI" dominated conversations. It highlights the fleeting nature of tech hype cycles and the eagerness to embrace the next "big thing." Agentic AI quickly became the most talked-about and simultaneously misunderstood technology, reflecting public fascination often detached from proper understanding.

This trend extends to the sudden reclassification of existing technologies as "AI" to capitalize on its marketing appeal. Tools and systems that have operated under different labels for years are now being rebranded as AI to ride the hype wave. For example, an advanced macro in Excel—once just a productivity feature—is now being called "AI-powered." Search engines remain fundamentally search engines, but some claim they have become "AI search assistants." Even automated call-routing systems, which have been around for decades, are now marketed as "AI-driven customer service platforms."

This dilution of the term "AI" misleads consumers and creates confusion about what true AI advancements entail. Companies risk undermining trust and devaluing genuine breakthroughs by labeling everything as AI.

In a moment that feels more like satire than an actual advertisement, Salesforce released "Rain," an ad that launched in December 2024 featuring Matthew McConaughey. It offers a hilariously bleak glimpse into the AI-driven future we supposedly need. McConaughey sits alone at an outdoor café in the pouring rain, lamenting how an AI-based app could have saved him from this damp inconvenience by automatically moving his reservation indoors. Across the street, his pal Woody Harrelson sits dry under an awning, beckoning him over. The implication? Without AI, we're hopelessly doomed to soggy dinners and missed opportunities.

But let's pause for a reality check. Are we now so lazy that we need AI to tell us it's raining? Are we so gullible to believe that restaurants wouldn't simply pull their tables inside? Is this ad reinforcing a narrative that human judgment is inherently flawed and requires technological intervention to function? What does the ad say about the priorities of tech companies? This ad epitomizes the confection of "needs" that don't exist, manufactured to sell a vision of technology solving problems consumers don't have. It's a glaring illustration of how hype often overtakes substance, perpetuating an AI ideology that inflates the role of technology in trivial aspects of life.

As Isaac Asimov once observed, "The saddest aspect of life right now is that science gathers knowledge faster than society gathers wisdom." This quote rings as accurate now as it did in 1988, reflecting the gap between technological advancement and our ability to integrate it wisely. Salesforce isn't just trying to sell AI; the company is offering a commentary on how easily we can be seduced by slogans and sales mantras, strategically aligned with a fear of missing out and ultimately closing our eyes to common sense.

Lost Art Forms

In November 2024, I came across a LinkedIn announcement by Andrew Chen, General Partner at a16z, heralding an investment in Promise AI, a studio claiming to be powered by "the world's best GenAI artists and storytellers." As someone with deep roots in media, entertainment, and gaming, my immediate reaction was one of disbelief. I couldn't help but comment, "This has to be a joke, right? It's not April Fools' Day, is it?"

What struck me wasn't just the hyperbole, but also the disconnection from reality. The announcement praised AI-driven storytelling as revolutionary, while overlooking the profound value of genuine human creativity, legal issues with copyrights and trademarks, and the complete disregard for how consumers want to be entertained. The comments section exploded with similar sentiments, condemning the notion that generative AI could replace or rival authentic talent. One user aptly wrote, "AI artists are not real. There is little to no talent behind the models borrowed, stolen, or modified." Another, with biting honesty, remarked, "'World's best GenAI artists and storytellers' is quite a statement. Also, I vomited while reading this post."

The hype surrounding such investments illustrates a troubling trend not exclusive to entertainment and gaming: the pursuit of technological spectacle at the expense of quality. This reckless prioritization of novelty over substance explains the layoffs, uninspired content, and failing connections with audiences that plague the entertainment world today.

I often think back to 1994 and wish I had proof of something I said while speaking at a classic film series showing Casablanca. I boldly told the audience, "There will come a time when we can put Humphrey Bogart, Marilyn Monroe, and Brad Pitt, into a movie together." At the time, it felt like a thrilling possibility. We've reached that point, and instead of excitement, it brings a sense of regret. The impact of our technological choices will forever shape how we view progress.

Coca-Cola's AI-generated holiday ads in 2024 illustrate the pitfalls of this approach. The campaign, intended to modernize the company's nostalgic "Holidays Are Coming" ads, used AI tools like Leonardo, Luma, and Kling to create visuals that include distorted proportions, glossy faces, and peculiar motion. Social media users quickly mocked these uncanny elements, with one calling the technology "a powerful tool" for producing flawed content. The backlash underscores the growing divide between cost-cutting executives embracing AI and creative professionals who argue it undermines artistry and quality.

Critics highlight deeper harms. AI-generated ads can mislead consumers with artificial imagery, damage brand trust, and exploit stolen creative work. Megan Cruz of The Broad Perspective Pod wrote: "This is always what [AI] was going to be used for btw. It's not some great equalizer. It's a way for already massively wealthy execs to add a few more mil to their annual bonuses by cutting creative teams entirely & having a machine vomit up the most boring slop imaginable instead."

Rob Wrubel, founder and managing partner at Silverside AI, which worked with Coke on the new campaign, said, "People believe that AI means you automate everything—someone just puts one prompt in, and a video pops out. It is the farthest from the truth," Wrubel said. But, he added, "The intensity of the creative process that used to happen over weeks and months can now happen every two hours."

While some studies suggest consumers may overlook AI flaws, such as Coca-Cola's nostalgic appeal masking errors, the broader implications remain troubling. AI's reliance on remixing preexisting content rather than creating genuine innovations erodes the integrity of brand storytelling. Brands risk alienating audiences and devaluing their identity when sacrificing quality for efficiency. This growing reliance on AI raises critical questions about authenticity, creative labor's future, and AI's ethics in advertising.

This "AI for the sake of AI" belief manifests as a costly distraction. Across multiple industries, executives are asking, "How do we take advantage of AI?" when the real question should be, "Do our customers want to engage with us through AI?" The distinction between these queries is not just semantic but foundational to building beneficial rather than merely fashionable strategies.

A compelling study in June 2024 (Cicek et al., 2024) offers a sobering insight into this issue. Their research, titled "Adverse impacts of revealing the presence of Artificial Intelligence in product and service descriptions on purchase intentions: the mediating role of emotional trust and the moderating role of perceived risk," sheds light on a profound consequence of misusing AI in consumer communication. The study presented 1,000 individuals with various product descriptions, some with AI mentions and others without. The findings were stark: products associated with AI were often met with decreased emotional trust, leading to lowered purchase intentions. This effect was exacerbated in high-risk scenarios, such as purchasing expensive electronics or medical devices, where the mention of AI features generated additional apprehension.

The repercussions of this bandwagon mentality are multifaceted. Companies risk not only financial waste from unwarranted investments in AI but also, perhaps more critically, the erosion of customer confidence. The allure of AI might bring initial attention, but if its application does not genuinely enhance the customer experience or offer a tangible improvement in product or service delivery, disappointment is inevitable. The chase after AI as a trendy tool rather than a strategic asset could lead companies into a cycle of unfulfilled promises and detached customer relationships.

It is wise for businesses to step back and evaluate their approach to AI. It isn't enough to plaster "AI-powered!" across marketing campaigns without substantiating how it tangibly benefits the customer. The focus should remain steadfast on the product's actual value and how AI can enhance that value meaningfully, not just as an "artificial" selling point.

During a CNN interview in August 2024, Gursoy elaborated on the study's findings, "We looked at vacuum cleaners, TVs, consumer services, health services, and in every single case, the intention to buy or use the product or service was significantly lower whenever we mentioned AI in the product description." This consistent trend across various industries underscores the depth of consumer skepticism toward AI-augmented products. Such views are further validated by recent studies, including a Pew Research survey of 11,000 people revealing a sharp decline in public trust toward AI: 52% of respondents in 2023 expressed more concern than excitement about AI technologies, a notable rise from 37% from the same research in 2021.

The reality? 99% of Americans already use AI-enabled products, yet 64% don't even realize it (Maese, 2025). At the same time, Gallup found that nearly three-quarters (72%) believe AI will worsen the spread of false information, and 60% fear it will shrink job opportunities. The research shows that the more AI integrates into consumers' lives, the more they distrust it.

Is it Hype?

In June 2024, the management consulting company Gartner reported that the industry surrounding Generative AI (GenAI) has reached the peak of inflated expectations and is now descending into the trough of disillusionment. This hype cycle describes a common pattern in technological adoption, where initial enthusiasm gives way to a more sobering reality.

The "peak of inflated expectations" occurs when product usage surges amidst significant hype and circumstance, yet tangible evidence that the technology can meet user needs remains sparse. Following this, the "trough of disillusionment" sets in as the initial excitement fades, early adopters begin to encounter performance issues, and the return on investment (ROI) appears to be lower than anticipated.

My perspective on this cycle, particularly with AI, suggests a divergence from typical Gartner cycles. AI technology, including its generative forms, has been evolving over decades, with foundational developments emerging as early as the 1960s. This long history challenges the notion of a sudden emergence typical of Gartner's "innovation trigger."

Throughout 2024, I've heard countless defenses and rhetoric urging us to "get used to AI; it's here to stay." However, in comparison, Virtual Reality (VR) has been around equally as long, but you don't hear the same demand for submission, mainly because the market size is still too small. Yet, the current trends in GenAI have seen dramatic increases in usage in less than two years, suggesting either a very slow disillusionment stage or an upcoming catastrophic crash akin to the dot-com bubble burst of the early 2000s. During that period, the NASDAQ composite stock market index soared by 800% between 1995 and its 2000 peak, only to plummet 78% by October 2002, erasing all its gains.

The parallel drawn here with the dot-com era highlights the potential volatility and the discrepancy between the long-standing development of AI technologies and the recent explosive hype. This mismatch between historical development and contemporary expectations suggests we might be on the brink of a significant recalibration in how GenAI is perceived and utilized in the broader tech landscape.

Widder and Hicks (2024) highlight that the deflation of the generative AI hype bubble reveals a recurring issue: technologies marketed as inevitable often leave behind harmful legacies. They argue,

> "Even as the generative AI hype bubble slowly deflates, its harmful effects will last: carbon can't be put back in the ground, workers continue to need to fend off AI's disciplining effects, and the poisonous effect on our information commons will be hard to undo."

Businesses rushing to adopt AI risk creating dependencies on flawed systems that fail to deliver on their promises. The authors warn that this hype-driven entrenchment damages customer trust, erodes the reliability of information ecosystems, and diverts resources from sustainable progress. They emphasize that, as AI's narrative shifts from inevitability to scrutiny, businesses must approach adoption cautiously, focusing on long-term utility rather than short-lived trends.

Both the hype cycle and bandwagon effect of rushing into AI can have severe repercussions for businesses. Beyond the financial implications of misdirected investments, there's a substantial risk of eroding customer trust and loyalty. History shows us that those chasing hype lose, while those who see through it build progress. AI will be no different.

The Make-Believe Technology

The British novelist Arthur C. Clarke famously said, "Any sufficiently advanced technology is indistinguishable from magic." Today, AI has become the epitome of this illusion, or what I often call, the make-believe technology.

In the 1970s, the Pet Rock became a cultural phenomenon. For $3.95, you could buy a rock in a cardboard box with breathing holes and a humorous training manual. It was absurd, it was viral, and it was ultimately empty. The Pet Rock didn't solve a problem or add lasting value—it was a gimmick packaged as a product.

In 2023, the belief in AI was steeped in doom, with references to *Terminator's* Skynet and fears of AI spiraling out of control to threaten humanity. This narrative dominated headlines. But in 2024, the doom perspective was conveniently dismissed, in part due to an essay by Marc Andreessen, who declared, "The era of Artificial Intelligence is here, and boy are people freaking out. Fortunately, I am here to bring the good news: AI will not destroy the world, and in fact may save it."

Andreessen's essay repeatedly couches AI as a utopian solution, ignoring risks and challenges while emphasizing limitless potential. This optimistic narrative reframed AI from a looming threat to a potential savior, perfectly aligning with Big Tech's agenda of rapid adoption and minimal regulation. By downplaying real risks and amplifying techno-optimism, the industry ensured that profits soared alongside AI models, creating a modern version of snake oil that many organizations still buy.

A Clearer Perspective

Let me be clear: AI itself is not the problem. The issue lies in the infallible ideology of AI, including rampant, misleading claims about its capabilities, harm caused by false promises, and reliance on stolen creative work to build these systems. There are seven main "types" of AI:

1. **Reactive Machine AI:** AI that reacts to external stimuli in real-time but lacks the ability to store information or learn from past experiences.
2. **Limited Memory AI:** AI that retains data temporarily to learn, adapt, and improve its performance for future tasks.
3. **Narrow AI:** AI programmed to perform specific tasks without the ability to independently learn beyond its initial programming.
4. **Theory of Mind AI:** AI designed to recognize and respond to human emotions while also performing functions similar to limited memory AI.
5. **Artificial General Intelligence (AGI):** AI designed to learn, reason, and perform tasks across various domains at a level comparable to human capabilities.
6. **Self-Aware AI:** AI that possesses self-awareness, understands human emotions, and demonstrates human-level intelligence, representing the highest stage of AI development.
7. **Artificial Superintelligence (ASI):** AI capable of exceeding human intelligence and outperforming humans in knowledge and capabilities.

We can find between 20-30 sub-forms of AI categorized under these main types. Currently, the most well-known Generative AI and large language models fall under Narrow AI because it is designed to perform specific tasks, such as creating text, images, music, or other forms of content based on patterns in the data it has been trained on. Strip away the jargon; their core function resembles a glorified search engine: prompt, response, repeat. Yet, the public fascination persists, driven by the desire to see AI as something more profound than it truly is. The reality starkly contrasts the *Star Trek* fantasies projected onto it. Instead of boldly venturing into uncharted territory, AI rehashes and regurgitates data. This book does not aim to marvel at the supposed "magic" of AI. Like the dystopian reality portrayed in *Wall-E*, it will explore how AI quietly erodes the depth of human intelligences, corrupting the qualities that make us uniquely capable of meaningful connection.

As we navigate the future, we see that aligning AI strategies with a nuanced understanding of customer behavior creates richer interactions and deeper engagement. But technology alone cannot replace the subtleties of human insight. So, we will also examine various types of intelligence—artificial, human, emotional, behavioral, and aspirational—unpacking how each plays a unique role in shaping customer experiences.

For this topic, I'll explore the groundbreaking work of Howard Gardner, a psychologist and professor of education at Harvard University, who introduced his theory of multiple intelligences in 1983. This influential framework redefined our understanding of intelligence by challenging the traditional focus on IQ as a single, standardized measure. Instead, Gardner proposed that intelligence encompasses a set of distinct abilities, each representing a unique way of perceiving, understanding, and interacting with the world. His theory highlights diverse forms of intelligence, from linguistic and logical-mathematical to spatial and interpersonal, providing a lens through which to contrast human capabilities with the functions of artificial intelligence.

Throughout this exploration, we'll consider how understanding customers' broader life goals and values can guide more mindful uses of AI. When businesses incorporate an understanding of these deeper motivations, they create strategies that not only fulfill immediate needs but also resonate with customers' long-term aspirations, enhancing engagement and loyalty. Balancing artificial intelligence with a genuine appreciation for human intelligence requires thoughtful strategy.

Artificial intelligence is impressive, scalable, tireless, and razor-focused. But for all its clever algorithms and data-crunching power, AI lacks the one element that human intelligence brings effortlessly to the table: **understanding**.

I've contemplated what I have to say about AI and its role in our lives and whether or not what is in my head is anything different from the hundreds of books on the same topic already available. To be clear: I'm not against AI; it can most definitely have a supportive place in our customer strategies. However, I will always favor focusing on the customer first and the technology last. Whether this is simply another vehicle to share fun stories, a genuine analysis of artificial hype, or a cautionary tale for where society is headed, in hindsight, I should have written this book sooner.

Chapter 1

A NEW HORIZON
Aspirational Intelligence in Customer Strategy

Aspirations are curious things. Although they often go unnoticed, hidden in the background of our personal and collective consciousness, they can be the driving force behind many of our most significant decisions. When aspirations align with the environment, including the ambitions of others, the outcomes can be truly transformative. Conversely, when aspirations clash, they can lead to frustration, resentment, and disillusionment.

The concept of "aspirational intelligence" captures the ability to understand, interpret, and act upon both personal and collective aspirations in ways that drive strategic decision making and innovative thinking. This form of intelligence seamlessly integrates emotional acuity with a forward-thinking approach, identifying and harnessing the motivations and desires that propel individuals and groups toward their future goals.

In the business realm, leveraging aspirational intelligence means tapping into the hopes of customers, employees, and stakeholders to design products, services, and experiences that satisfy current needs and resonate with their future ambitions. This strategy demands a profound understanding of what truly drives people, encompassing their dreams and enables organizations to forge deeper, more meaningful engagements. Further, it guides companies in setting visionary goals and developing strategies that deeply connect with their target audience, fostering loyalty and driving sustainable growth.

Expanding on the concept of aspirational intelligence as it relates to the AI field, we see that AI's capabilities can extend beyond mere data processing and task execution based on predefined algorithms. In fact, AI can also support long-term goals or missions aligned with broader human values or business objectives. This advanced form of intelligence can support bridging the gap between current capabilities and visionary goals, ensuring that decision-making processes are not only efficient but also ethically sound and conducive to future growth.

In 2021, at Women's Wear Daily's (WWD) Digital Beauty Forum, Jane Lauder, chief data officer at The Estée Lauder Companies Inc., introduced the concept of "aspirational intelligence" as a cornerstone of their data strategy, describing it as "connecting the dots to find something desirable and aspirational for consumers." This concept underpins the company's efforts to leverage data in innovative ways to anticipate and meet consumer desires. Lauder elaborated on her vision by sharing the development of an "infinity loop" framework designed to streamline information-sharing across the organization. At the heart of this loop lies data, serving as the engine driving operational efficiency and aspirational consumer engagement.

At the Cannes Lions 2024 panel, Ayami Nakao, chief

strategy officer at Hakuhodo, a retailer of fine makeup brushes, emphasized the holistic nature of consumers, rooted in the traditional Japanese concept of sei-katsu-sha. This perspective views consumers not just as buyers but as whole individuals with complex desires and needs. The panel discussed three aspirational pillars: health and wellness, success and happiness, and love and intimacy, demonstrating how advertising can challenge and positively influence cultural aspirations.

At Axios' What's Next Summit in 2024, Reshma Saujani, founder of Girls Who Code and Moms First, further championed the role of AI in addressing significant societal challenges. She advocated for "aspirational AI, not just ethical AI," highlighting the importance of inspiring and including underrepresented groups in technology and empowering individuals to create, build, and innovate.

This dialogue around "aspirational intelligence" and AI's potential within the space raises compelling questions about the nature of AI recognizing individual desires. Can AI possess desires or ambitions akin to humans? In Philip K. Dick's *Do Androids Dream of Electric Sheep?*, which inspired the film *Blade Runner*, replicants struggle with their own identities and aspirations, blurring the lines between human and machine desires. Films like *Ex Machina* and *A.I. Artificial Intelligence* further explores this theme, depicting AI entities seeking purpose beyond their programmed existence.

The quest for AI aspirations parallels the evolution of customer expectations in today's market. Just as AI characters in fiction strive for meaning, customers seek transformative experiences that fulfill deeper desires. Companies like Apple and Nike excel in this area, selling not merely products but also visions of innovation and empowerment that resonate deeply with consumer aspirations. This synergy between AI's quest for self-actualization and customers' pursuit of transformative experiences underscores a shared narrative of striving for greater significance, a narrative that connects the aspirations driving both fictional AI and real-world consumers to a universal desire for growth, meaning, and connection.

Understanding these aspirations, whether it's the literal quest of AI for greater autonomy or the consumer's search for meaningful engagement with brands, reveals a fundamental business truth. The success of our strategies hinges not merely on meeting expectations but on aligning with the deeper aspirations of all stakeholders involved. As we delve deeper into customers and workplace culture dynamics in the chapters ahead, I aim to illustrate that aspirations are the engine of business growth. Effective leaders must recognize and harness these aspirations not as an afterthought but as a cornerstone of contemporary customer strategy and customer-centric cultures.

The Workshop That Never Was

Workshops aren't really about technology. They're about purpose, about connecting each tool and tactic to the core reason a business exists: to better serve its people, to better serve its customers. I learned this powerful lesson leading a session for Ferguson, the largest U.S. distributor of plumbing supplies, HVAC equipment, and industrial products.

Our team had crafted a four-day, technical workshop around integrations leveraging Application Programming Interfaces (APIs), AI applications, and security systems. It was meticulously planned to cover every latest advancement that was relevant to their business. But on the first day, I sensed a disconnect. The room was quiet, and not in a focused way. Even as we moved through the carefully curated agenda, I could feel the disengagement. The slides, the demos, they weren't sparking the enthusiasm we'd hoped for. It began to dawn on me that maybe we'd missed the real purpose of our time there.

While mentally reviewing the day that evening in my hotel room, I realized what we'd overlooked: our purpose with this workshop wasn't about technology. This team hadn't gathered to learn tools; they wanted to talk about aspirations—for better customer service, for a more stream-lined business, for systems that could bring them closer to their customers. They weren't asking, *How does this tech work?* They were asking, *How can this technology help us become indispensable to our customers?* What they wanted was a deeper dive into customer strategy.

The next morning, I stepped in front of the group and took a different approach. "Let's start fresh," I said, dropping the agenda we had prepared. "Forget the topics. Let's talk about what's important to you right now." I grabbed a marker and wrote on the board, "What would make the biggest difference for your customers?"

Hands went up immediately, the energy shifted, and the room came alive. As we shaped a new agenda together, themes emerged around customer strategy and the power of aspirations. They wanted to know how the tools could empower them to serve more effectively, solve customer issues faster, and anticipate their needs before they arose.

From that point on, the workshop transformed. No longer a structured series of technical demos, it became a collaborative experience that aligned with Ferguson's higher purpose. Each session felt dynamic, connected to their aspirations and grounded in what they truly wanted: to be the best possible partner for their customers. What started as a technical workshop had evolved into something much more: a roadmap for achieving aspirations.

Here's what this experience taught me: customer strategy is inseparable from aspirational intelligence. Aspiration is the fuel that makes every decision meaningful. It's the invisible force that turns technical know-how into a path forward, that moves an organization beyond mere transactions into creating real value. By daring to ask, *What do you aspire to achieve?* and then listening deeply, we unlocked the workshop's potential to become what it needed to be all along.

In the end, it wasn't just a workshop. It was a blueprint for how Ferguson could meet the future with purpose. And it wasn't just beneficial; it was aspirational. Because when you connect each tool, each decision, and each strategy to the power of aspirations, you don't just lead a workshop— you build a partnership driven by the goal of delivering real value to the people who matter most.

The Power of Aspirations

First, let's clear up potential confusion: When we speak of "aspirations" in this present context, we are not equating them with the generally unrealistic, unattainable lifestyle options so frequently dangled before our eyes by media sources. Although it is certainly possible to aspire to the "lifestyles of the rich and famous," the aspirations in focus here are typically more mundane but also (for this same reason) more achievable aspirations that inspire us in our daily personal and business lives.

Along these lines, the American Psychological Association defines an aspiration simply as "an ambition, goal, or any kind of desired end that might be achieved through personal effort." Aspirations in this sense may be considered roughly synonymous with "goals," although, as psychology writer Kendra Cherry (2024) explains, goals "tend to be backed by actions and are often centered on the short-term or near future" whereas aspirations "tend to be much more future-focused and are often more general."

Cherry goes on to note the major impact aspirations have on our lives: They can both "help you make life choices" and also steer you to "engage in certain behaviors that can put you on the path toward achieving those life aspirations." In this way, aspirations open a new horizon of possibilities, helping us feel more motivated and inspired. When you have an image of what you want in your mind, it can help you stay focused on the things that you need to do to ultimately make that dream happen. This focus and inspiration can provide the drive we need to overcome the obstacles that often arise on our personal growth or professional development journey. As the Berkeley Well-Being Institute (2022) observes, our aspirations also reveal what we as individuals or a society value most. "If you aspire to be a good parent, that means your family is important to you. If a society has aspirations of ending homelessness, that means it values and respects its citizens," (Hart, 2016). By motivating us to work for the realization of our values, aspirations give us a sense of purpose, meaning, and accomplishment.

At its deepest level, then, the concept of aspiration is about more than simply achieving or gaining certain outcomes: It is both reflective and constitutive of our identity. As famed Lebanese-American poet Kahlil Gibran once said, "To understand the heart and mind of a person, look not at what he has already achieved, but at what he aspires to" (Gillett, 2019). Our aspirations in this way serve as a window to our true selves or the selves that we wish to become. In the words of Courtney Headley (2023), global head of culture and inclusion at Quest Global, aspiration is "a fuel that ignites the fire within, propelling us toward our dreams and unlocking our true potential." For most (all?) of us, there is a gap "between where [we] are now and where [we] need to be" to achieve our fully actualized selves. "Aspiration," Headley suggests, "represents the drive to bridge that gap."

The power of aspiration as a motivator to action and self-actualization was brilliantly embodied by Nike's famous "Just Do It" campaign launched in 1988. Created by Dan Wieden of the advertising firm Wieden+Kennedy, the tagline was meant to "help expand the Nike brand's consumer base beyond hardcore male athletes," (AAAA, 2017). As Nike Co-Founder Bill Bowerman put it at the time, "Everybody has a body and is therefore a potential athlete." Clearly the message struck a chord, as the ad "became a universal rallying cry that resonated both inside and outside of sports, for generations to come." The universality of the ad's appeal attests to the universality of aspiration as a powerful actualizing force in human life.

Three Types of Aspiration

So far we have been speaking of "aspiration" as a more or less unidimensional phenomenon, but it turns out that there are various ways of classifying aspirations. One way commonly used by researchers is to distinguish between *intrinsic* and *extrinsic* aspirations. Cherry (2024) explains,

> the former are meant to "help satisfy psychological needs" and enhance "an individual's well-being"; these aspirations include such aims as building relation-

ships, contributing to one's community, having children, and developing one's skills and talents. Extrinsic aspirations, in contrast, are "focused on achieving an instrumental outcome" such as achieving wealth or popularity.

Although we may draw at times on this and other ways of classifying aspirations over the course of this book, let me offer my own rather different threefold categorization of aspirations I believe will serve as a more pertinent point of reference for the discussions that follow:

Personal Aspirations – These are individually held aspirations intended primarily for the individual's own benefit. You may have an aspiration to improve your health by exercising regularly and cutting down on processed foods, or you may be thinking of furthering your education and earning a graduate degree.

Granted, accomplishing these goals is almost certain to have various impacts—both intended and unintended—on others as well. Being healthier allows you to live a more active and potentially longer life, which affects your family and friends and those who benefit from your professional efforts. At their core, these are personal decisions you are making largely for your own benefit.

Group Aspirations – These refer to aspirations that are shared by the members of a group readily capable of acting in concert, whether an organization, association, business, or any other society of individuals. In the most typical cases, these aspirations are held primarily for the benefit of the group itself and its members. You might spend every Thursday night knocking down pins with fellow members of a bowling team whose aspiration is to attain the #1 position in its league.

Again, the team's aspiration to win will likely impact not only the group but others as well, from individuals such as spectators, to other groups like competing teams, to entities larger than the team itself, including the league. Group aspirations tend to predominantly have the benefit of the group itself in view. Because the aspiration is typically shared among the members of the group, a mutually reinforcing synergy may develop that increases the likelihood of the aspiration being fulfilled, a "teamwork" effect. Note that the size of the group in question may range from a minimum of just a few individuals, such as a married couple or founding partners of a startup, up to any larger collective capable of working or striving together toward a common goal, like a company.

Global Aspirations – These aspirations have in view an even broader, more complex range of actors and beneficiaries. Global aspirations are not limited to groups or associations of individuals readily capable of operating in concert (as with group aspirations) but can also extend to much more complex and multi-faceted conglomerates and causes. Multinational corporations or consortia of businesses or agencies, broad political entities such as cities or nations, and stakeholders in issues aiming to benefit a broad slice of human society or even the entire human race, such as environmental concerns, may all hold global aspirations.

For a relatively straightforward non-business example of the above distinctions, consider an individual's relation to religion. A religious person might have a personal aspiration to embody their religious beliefs sincerely through their personal choices and actions. At the group level, the same person may be involved in a local assembly for collective worship, receiving or giving religious instruction, and similar activities. The individual's group aspirations to be part of a thriving congregation would likely be shared by many other congregants and would incorporate consideration of the good of the whole, not simply of the individual. Finally, that same local assembly might be associated with a much larger entity such as the Catholic Church, which includes many people the individual and other members of the local assembly would never meet or consciously coordinate with. Their aspirations for this larger organization, such as functioning as an instrument for God in the world, could be considered global in nature.

In a business context, I might choose to work with a particular company, hope to receive a promotion after time served, and desire to work on a specific project. These would all be *personal* aspirations. Along with the other members of my team, I may aim to be #1 in sales and make a notable positive impact on the company's bottom line. These would be *group* aspirations. Finally, I may join with other colleagues in supporting the company's push to become sustainable and make a broader positive impact in our community and on the environment. These could be considered *global* aspirations. This business perspective is relevant, where leadership direction and company culture often evolve over time.

Aligning the Strategies

Integrating aspirational intelligence as a foundational customer strategy is essential as it aligns with how personal aspirations evolve and how business goals must also adapt to enhance customer loyalty, employee engagement, and overall organizational success. There is compelling evidence that leveraging aspirational intelligence in this manner is becoming a central decision point for customer success. According to a Deloitte report referencing the 2021 Edelman Trust Barometer, 68% of consumers feel empowered to demand change within corporations, and 86% expect CEOs to actively address societal issues. This trend underscores the importance for businesses to resonate with both the social and global aspirations of their customers.

CEOs are increasingly recognizing societal impact as a key performance indicator and are steering their businesses toward a socially conscious ethos. Research from the Boston Consulting Group (BCG) supports this shift, showing that investments in sustainability issues material to a business can enhance shareholder returns by up to 5% (Hutchinson et al., 2024). Conversely, superficial investments may reduce returns by as much as 2%. My own data indicate a 300% increase in the likelihood of successful business transformation when performance metrics align with aspirational metrics. This data underscores the value of applying insights from the aspirational intelligence within a business context.

The alignment of customer strategies with a broad array of stakeholder concerns offers rich potential for business leaders. Challenges remain, as evidenced by findings from Achievers, where the top factor of employee satisfaction is the alignment of values with an organization. "Over 75% of employees consider it very important to work for a company with defined core values. This translates to a better bottom line: companies with highly aligned cultures see 30% higher growth" (Wong, 2024). This stat demonstrates significant gaps between employees' aspirations and the company's goals, highlighting the urgency of intentionally and methodically aligning the two.

By acknowledging and strategically responding to these core human desires through aspirational intelligence, businesses can significantly strengthen customer loyalty, increase employee engagement, and drive overall organizational success. This approach not only meets the immediate needs of individuals but also aligns with broader societal values, creating a sustainable path forward for contemporary businesses.

Understanding Predictive AI

Predictive AI focuses on using historical data to anticipate future outcomes. It takes patterns—what happened yesterday and today—and analyzes them to forecast what will happen tomorrow. Think of it as a tool that turns raw information into actionable insights. Unlike Generative AI, which creates new content like text, images, or designs, Predictive AI focuses on projection. Where Generative AI answers creative questions such as, "What could exist?" Predictive AI tackles strategic ones, like "What is likely to happen, and how can we prepare?"

Businesses often look to Predictive AI to stay ahead of their competition. It can identify customer behavior trends, predict market changes, and flag potential issues before they arise. Imagine knowing a customer might stop purchasing or that a market segment is about to shift, all before these events happen. Predictive AI allows businesses to move from reacting to being proactive.

However, Predictive AI is not without its limitations. It relies on historical data, which may not always reflect emerging trends or shifts in human behavior. Consumer preferences can change suddenly due to societal, cultural, or economic factors not captured in historical datasets. Additionally, Predictive AI struggles with entirely new scenarios where no relevant historical data exists. In such cases, relying too heavily on predictions can lead to misguided strategies and missed opportunities.

This is where Aspirational Intelligence offers a contrasting perspective. While Predictive AI anticipates what is likely based on past patterns, Aspirational Intelligence focuses on understanding deeper human motivations and desires. Predictive AI might tell you that a customer will likely buy a particular product. Aspirational Intelligence asks why they want it and what it says about their broader goals and values. It shifts the focus from probability to purpose, providing insights that can guide the next move and the overall vision.

Using Predictive AI

For businesses, the potential of Predictive AI is undeniable. It helps refine customer strategies by identifying patterns in purchasing behavior, preferences, and engagement. A retailer might use Predictive AI to stock inventory based on anticipated demand, minimizing waste and maximizing profit. It enables organizations to personalize customer experiences at scale, crafting interactions that feel thoughtful and intentional.

Yet, Predictive AI must be applied carefully. Over-reliance on predictions can alienate customers if the insights feel overly intrusive or inaccurate. Sending targeted ads based on poorly understood patterns may create the impression of surveillance rather than service. Ethical considerations also come into play, especially when using Predictive AI to influence sensitive decisions, such as credit approvals or healthcare recommendations. Transparency in making predictions and ensuring fairness in their application are critical to maintaining trust.

One area where Predictive AI has potential is cybersecurity. Government agencies are increasingly leveraging this technology to bolster zero-trust architectures, as highlighted by Stephen Wallace, CTO of the Defense Information Systems Agency, during the 2024 Rubrik Public Sector Summit. Predictive AI's ability to analyze large datasets allows it to detect early warning signs of ransomware or mass encryption events, which traditional endpoint security measures might miss. Travis Rosiek states, "Predictive AI can strengthen agencies' cybersecurity resilience and will likely become more important than generative AI because larger data sets will be at its disposal to anticipate attacks and trends" (Rosiek, 2024).

By integrating predictive AI with behavioral analytics, agencies can identify anomalies such as unauthorized access or privilege escalations before they escalate into significant threats. This proactive approach strengthens cyber resilience and supports the broader mission of protecting national security and ensuring uninterrupted service delivery.

The Balanced Approach

The key to leveraging Predictive AI effectively lies in understanding its role as a tool rather than a standalone solution. Predictive AI provides invaluable insights into what might happen next, enabling businesses to anticipate needs and refine their strategies. However, it is not a crystal ball and should not be used to replace human judgment or strategic foresight.

An easy way to think of Predictive AI is to compare it to predictive text features on your phone or autocomplete in applications like Gmail. These systems predict what you might want to say based on patterns observed in what others have written in similar contexts. While this technology often feels intuitive and valuable, it is far from perfect.

A popular example is social media trends, where people start a sentence and allow predictive text to be completed by consistently selecting one of the system's suggested words. For instance, starting with "I was..." might generate suggestions like "going," "looking," or "there." For this exercise, I selected the middle option 15 times, resulting in the sentence, "I was looking a lot like a new branch and a genie for a few years now." The outcome is often humorous and nonsensical—highlighting that Predictive AI, without the nuance of human oversight, risks becoming a sophisticated version of "Mad Libs."

This demonstrates the limitations of Predictive AI when used in isolation. While it excels at identifying patterns and trends, it lacks the deeper context, intuition, and creativity humans bring. For businesses, Predictive AI should be viewed as an augmentative tool that can enhance decision-making but requires human intelligence to validate, interpret, and act on its suggestions. Together, AI and human insight can create a balance that delivers practical, meaningful results.

By combining Predictive AI with Aspirational Intelligence, businesses can address both the tangible and intangible aspects of their strategies. Predictive AI answers the question, "What can we expect?" Aspirational Intelligence pushes further, asking, "What should we aim for?" Together, these approaches create a framework that balances foresight with purpose, helping businesses thrive in a constantly evolving landscape.

Aspirational Intelligence vs Predictive AI

Aspirational intelligence equips humans to dream, innovate, and imagine possibilities that transcend existing norms. This unique ability enables businesses to craft transformative strategies that redefine industries by focusing on creative, forward-thinking solutions to customer needs. Predictive AI excels at analyzing historical and real-time data to forecast trends, identify customer preferences, and assess risks. It provides a strong foundation for decision-making but lacks the creativity and vision to conceptualize novel strategies. Predictive AI relies on established patterns, making it less effective in navigating disruptive or unforeseen market shifts.

Aspirational Intelligence and Predictive AI for Customer Strategy

- **Aspirational Frameworks:** Use predictive AI to map potential customer aspirations, but rely on human insight to construct emotionally resonant strategies that transform aspirations into actionable opportunities.
- **Visionary Gap Analysis:** Leverage predictive AI to identify gaps between customer behaviors and unspoken aspirations. Human intelligence bridges this gap by crafting emotionally meaningful pathways to future customer needs.
- **Scenario Co-Creation:** Use AI to outline likely scenarios, then employ human imagination to explore aspirational "what if" possibilities that challenge conventional outcomes and create innovative customer strategies.

Chapter 2

BEYOND OUR LIMITS
Logical Intelligence in Customer Insights

I was asked to speak at an AI conference in Miami in the summer of 2024. After my presentation, I was pulled aside by the vice president of marketing for a mid-size commerce company. Sarah had been eagerly awaiting insights from the company's newly implemented AI tool that her team promised would revolutionize their approach to customer segmentation, allowing them to tailor messaging with pinpoint precision. The results had come in over the weekend, and she was practically glowing with enthusiasm while asking if I would be available to review them.

"I think we're about to hit the jackpot," she whispered to me. "This is exactly what we've been looking for! The AI has segmented our customers into these perfect groups, and it's showing us what we've been doing wrong all along. Could you take a quick look and confirm what I'm seeing before we move forward?"

Her excitement was contagious, and I couldn't help but feel curious. AI tools can accomplish a lot, no doubt, but they're only as good as the data they're trained on and the algorithms guiding them. Sarah had been under pressure to deliver results that could reshape the company's marketing strategy, so I could see why she was thrilled to have something tangible to work with.

I reviewed the data the following week, and on the surface it looked perfect. The AI had neatly categorized the company's customers into five distinct personas, each with its own detailed profile: shopping habits, preferred products, and even suggested marketing messages. It appeared to be the kind of insight every marketing VP dreams of—clean, actionable, and data-driven.

But something was off. It was *too* perfect.

As I began digging into the data, I noticed subtle inconsistencies. The segments didn't quite align with data from previous campaigns. Certain customer profiles seemed over-exaggerated with percentages of improvement, as though the AI had stretched its analysis to create a compelling narrative. Personally, it often feels as if AI wants to appease us. One persona was labeled as "Loyalists" a group of supposedly long-term, high-spending customers. But when I checked the raw data, this group was filled with customers who had only made one purchase. The AI had misinterpreted their engagement, categorizing them based on vague loyalty metrics.

Then I noticed something more alarming: some of the key insights, such as buying motivations or customer sentiment, were completely fabricated. I cross-referenced this with the data the AI was supposedly drawing from, and it became clear that it had "hallucinated" these insights. There was no actual evidence to back up some of the claims it was making.

It wasn't that the AI had created completely false data but, rather, it had woven a narrative from fragments of information that didn't hold up under scrutiny. And worse, it was exactly what Sarah was hoping to see. Her marketing team had been struggling for months to justify their current strategy, and now this AI-generated report conveniently confirmed that they were on the right track, all wrapped in data that looked real and actionable. It was easy to see why Sarah was eager to move forward with it.

AI's Role in Collecting and Analyzing Customer Data
The internet is becoming increasingly cluttered with low-quality AI-generated content, a trend that is accelerating at an alarming rate. A recent study from researchers at Amazon Web Services (AWS) AI Lab highlights this troubling shift, revealing that more than half the sentences on the web, 57.1% have been translated into two or more languages, likely using Large Language Models (LLMs). The poor quality of these translations, especially in lower-resource languages, suggests widespread reliance on AI models for content generation and translation. This influx of synthetically created content, being referred to as de-generative AI, diminishes the quality of online information, making it harder for users to discern reliable sources. Europol's recent report adds further urgency, warning that by 2026, up to 90% of online content could be AI-generated, raising concerns about disinformation and the overall reliability of what we encounter on the internet (Dupré, 2022). As AI-generated content proliferates, the need for critical evaluation of online information is more pressing than ever.

One flaw with the influx of content is that it has turned AI into a data cannibal, an issue that occurs when AI-generated content circulates back into the training datasets used by other AI systems. As the internet becomes flooded with AI-generated articles, translations, and images, these outputs are often scraped and fed back into new AI models for further learning. Instead of rich, human-created data, these models learn from their own recycled outputs, leading to a feedback loop of increasingly homogenized, inaccurate, or biased information. This self-referential loop reduces the diversity and originality of the content that AI models rely on for training, which in turn lowers the accuracy and usefulness of the models' outputs.

For customer insights, this is particularly concerning because businesses depend on AI to generate actionable, accurate data about customer behavior and preferences. If AI is learning from AI-generated content rather than from real-world human behavior, the resulting insights can become less representative of actual customer needs. This can lead to faulty marketing strategies, product misalignment, and poor decision-making. AI models trained on inaccurate or repetitive content risk creating a distorted view of consumer trends, which can undermine trust and prevent businesses from responding effectively to their customers' true preferences and behaviors. Ultimately, the value of AI in understanding customers diminishes as the quality of its underlying data deteriorates.

With AI, data collection has scaled to a level that would have been unimaginable just a few years ago. Modern AI systems sift through millions—sometimes billions—of data points, synthesizing them into comprehensive reports that promise to offer deeper insights into customer behavior than ever before. The idea is tempting: that AI can assemble a complete, nuanced picture of a customer without requiring human involvement.

It reminds me of the Mr. DNA scene in the 1993 film *Jurassic Park*.

"Thinking machine supercomputers and gene sequencers break down the strand in minutes. And virtual-reality displays show our geneticists the gaps in the DNA sequence. We used the complete DNA of a frog to fill in the holes and complete the code."

But here's where the allure of AI often overshadows reality. The promise of AI-driven data analysis rests on two pillars: the quality of the data it receives and the sophistication of the algorithms processing that data. AI is essentially a reflection of the information it is fed, and as the saying goes, "garbage in, garbage out." If the data is biased, incomplete, or outdated, the insights generated by AI will be flawed.

Philosopher Shannon Vallor, in her book *The AI Mirror*, uses the metaphor of a mirror to critique how artificial intelligence reflects human data and biases. Inspired by Chicago's iconic Bean sculpture, Vallor observed how its surface distorts reflections, much like AI mirrors human inputs without objectivity. Unlike metaphors that liken AI to sentient entities, Vallor argues that AI is more akin to a "flat, inert surface," captivating us with illusions of depth but lacking true understanding. She draws parallels to Narcissus from Greek mythology, warning that humanity risks losing itself to the reflection AI provides. While Vallor acknowledges AI's potential in solving complex health and environmental challenges, she describes her relationship with technology as one that has "slowly turned sour," reflecting her concern about the societal consequences of unchecked AI growth.

Vallor emphasizes the importance of cultivating human virtues such as justice and wisdom to navigate our evolving relationship with AI. She warns that AI, trained on historical data, reinforces existing societal issues like racism, poverty, and inequality, thereby limiting our ability to address novel challenges. AI mirrors only past human behaviors, offering no insight into emergent problems or moral growth. Vallor cautions that over-reliance on machines optimized for efficiency and profit could erode the values that make life meaningful. Instead of seeing AI as an opponent to be defeated, she urges us to focus on maintaining our uniquely human capacities, such as moral judgment and creative thought. "We don't need to 'defeat' AI," she asserts. "We need to not defeat ourselves."

AI doesn't have the ability to filter out bias or interpret missing data; it works with what it has, leading to a mirror that "sees" the customer. But in reality, what it knows is only a small portion of the entire experience.

In a business scenario, AI might track that a customer abandoned their cart during an online purchase. The data points could show that the customer spent a significant amount of time browsing products, added several items to their cart, but then suddenly left the site without completing the purchase. Based on these observations, AI might draw conclusions about the customer's likelihood to buy in the future or suggest retargeting them with promotions. However, the AI cannot discern the underlying reasons for the abandonment. Maybe the customer was interrupted by a phone call, experienced frustration with the website's functionality, or decided to delay the purchase. The algorithm can only process the behavior, not the emotional or situational context behind it.

This gap between data-driven insights and emotional or contextual understanding is where AI often falls short. According to Haan and Watts (2023) in a Forbes article, more than 75% of consumers are concerned about misinformation from artificial intelligence. Many customers are not even aware of the extent to which their data is being collected and analyzed.

Recent legal challenges, such as the class-action lawsuit against General Motors (GM), reveal profound privacy concerns over the use of connected vehicle data. This lawsuit highlights how GM shared sensitive driving data with insurance companies, leading to unexpected rate hikes for drivers, a clear breach of trust. Furthermore, Texas Attorney General Ken Paxton sued GM for misleading Texans into sharing their data, which was then sold to data brokers. As Paxton stated, "General Motors deceptively collected scores of data points from consumers about their driving habits, monetized that data by selling it to other commercial actors, and permitted those actors to use the ill-gotten data to make adverse decisions when dealing with those same consumers" (Ng, 2024). These cases underscore the urgent need for enhanced regulatory oversight and transparency in how companies collect and use data, stressing the ethical responsibilities businesses hold in managing personal information in the age of AI.

Several studies highlight a significant trend among consumers who are increasingly willing to share personal data in exchange for customized experiences and incentives. According to industry newsletter *Retail Brew*, 73% of US respondents are prepared to share their data for deals, with 65% extremely likely to opt-in for exclusive sales. The Trade

Desk Intelligence and YouGov study in 2023 found that 74% of American adults are also inclined to share personal information for deals and perks (Clark, 2024). This openness reflects a growing consumer acceptance of a value exchange where personalization and targeted content, such as recommendations based on past behaviors and location-based offers, are provided in return for their data. However, this willingness exists alongside growing concerns about privacy, highlighting the personalization–privacy paradox.

The personalization–privacy paradox reflects the tension between consumers' appreciation of tailored experiences and their concerns over privacy risks (Cloarec, 2020). While personalization adds value by enhancing convenience and relevance, the intrusive nature of data collection often erodes trust. Cloarec (2020) notes that advanced anonymization technologies, though effective, often prove too complex for average users. This paradox is compounded by the "attention economy," where firms like FAANG (Facebook, Amazon, Apple, Netflix, and Google) leverage consumer data to capture attention and market it to advertisers. Resolving this requires rethinking personalization through choice architectures and transparency, ensuring data practices align with user expectations and ethical standards.

The chart on the right illustrates this dynamic by categorizing users into four quadrants based on their perceptions of personalization's value and privacy risks. This framework reveals how user preferences and trust impact their data-sharing behaviors, emphasizing that decisions often reflect personal priorities, not strict logic.

PERSONALIZATION-PRIVACY PARADOX MATRIX

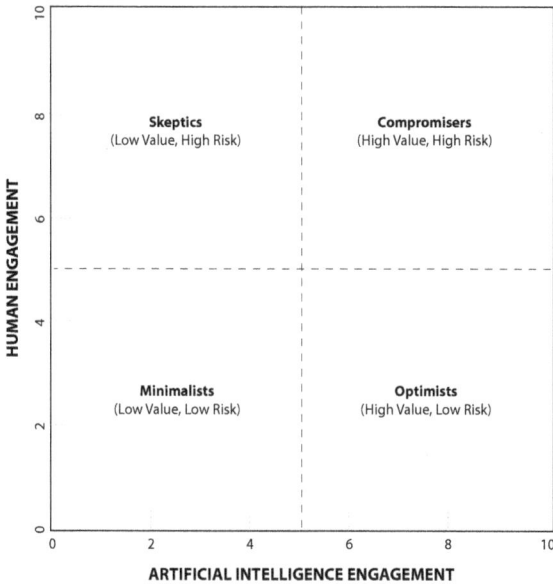

1. Minimalists: These individuals value their privacy highly and see little benefit in personalization, opting for generic, low-risk services.

2. Optimists: They perceive high value in personalization while trusting platforms to maintain privacy, fostering greater engagement.

3. Skeptics: Skeptical of both privacy and personalization, these users avoid data sharing, doubting its utility.

4. Compromisers: They recognize significant privacy risks but willingly trade them for valuable, highly tailored experiences.

What is Analytical AI?

Analytical AI refers to systems that process complex data, identify patterns, and provide actionable insights through logical and structured analysis. It dissects existing information to draw conclusions, solve problems, and optimize decisions.

At its core, Analytical AI operates through algorithms and machine learning models that are highly effective at identifying correlations and causations within large datasets. These systems can evaluate historical data, detect trends, and forecast future outcomes with remarkable precision. Analytical AI excels in financial analysis, healthcare diagnostics, marketing optimization, and customer behavior insights, offering businesses a deeper understanding of their operations and markets.

One of its primary strengths lies in its ability to process diverse datasets, including structured, unstructured, and real-time data. By leveraging techniques like natural language processing (NLP), predictive modeling, and statistical analysis, Analytical AI can uncover insights that traditional methods might miss.

However, the effectiveness of Analytical AI is deeply tied to the quality of its data. When trained on biased or incomplete datasets, even the most advanced Analytical AI systems can perpetuate existing inequities or produce skewed results. This highlights the importance of addressing bias in data to ensure fair, accurate, and inclusive decision-making.

Navigating the Trap of Degenerative AI

In the rush to leverage AI, many businesses are falling into a dangerous trap: degenerative AI. Unlike AI systems designed to optimize and add value, degenerative AI refers to the risk where AI models lose accuracy and effectiveness over time. This decline can occur when models are trained on their own outputs or rely on data that's flawed or contaminated, leading to increasingly inaccurate results. This decay can also occur when the models are trained using fake news, spam, or more generally, "garbage."

The trap not only hampers progress but also limits the potential to move beyond our limits in understanding customer behavior.

It works against customer-centric insights, leading to skewed accuracy and hollow results. It's a reliance on feedback loops that reinforce biases, distort perceptions, and degrade our understanding of customers. And when we lose the nuance of our customer base, our ability to innovate and respond authentically diminishes.

Degenerative AI systems often perpetuate existing patterns, replicating rather than questioning. They rely on datasets riddled with biases or narrowly focus on historical data without accounting for emerging needs. The result is a system that lacks the capacity for true adaptability.

In August 2024, Aatish Bhatia, a writer for the New York Times, interviewed AI researchers and fed an AI system its own output. One test featured AI-generated digits, which initially replicated handwritten numbers with great accuracy but degraded over successive generations, ultimately converging into indistinct shapes. This phenomenon, known as "model collapse," exemplifies the diminishing quality of AI systems trained on their output (Bhatia, 2024). As Bhatia observes, this degeneration underscores the necessity of "high-quality, diverse data" to preserve the integrity and performance of AI models. Without such input, these systems risk producing progressively narrower and less effective outputs.

	"3"	"4"	"6"	"8"	"9"
Handwritten digits	3	4	6	8	9
Initial A.I. output	3	4	6	8	9
After 10 generations	3	4	6	8	9
After 20 generations	3	4	6	8	9
After 30 generations					

Aatish Bhatia (2024)

The implications of this trend extend beyond theoretical models. While synthetic data can supplement training in specific contexts, Bhatia (2024) emphasizes that "there's no replacement for the real thing." For AI systems to sustain their effectiveness and relevance, researchers and developers must prioritize fresh, diverse datasets in their training pipelines. This strategy safeguards against the collapse of quality and diversity, ensuring AI models evolve robustly and remain aligned with their intended purposes.

Degenerative AI also carries significant implications for customer insights. Without authentic and diverse data, customer insights systems risk delivering outputs that reflect narrowing perspectives and generating dimensioning representations of real customers. This creates a dangerous cycle for businesses: relying on a feedback loop that fails to adapt to customers' deeper, evolving needs.

Take my consultation story in the chapter opening. If the AI only analyzes past sales data, it might suggest that trends from last season will work again next season simply because products sold well before. Analytical AI, however, offers a crucial distinction. It dives deeper, interpreting nuanced shifts in customer desires, values like sustainability, unique self-expression, and even subtle cues from changing social contexts. While generative AI explores possibilities by synthesizing and expanding upon data, analytical AI identifies patterns and trends even when historical precedents are lacking. This forward-looking lens uncovers insights beyond intuition or existing information, equipping businesses to understand and anticipate their customers' needs truly.

What's the alternative? Generative AI fosters creative exploration, enabling businesses to test hypotheses and imagine customer behaviors in new ways. Analytical AI, on the other hand, brings a disciplined focus, uncovering underlying patterns and relationships that remain hidden in complex datasets. Together, they form a powerful duo: generative AI inspires broader thinking, while analytical AI grounds it in evidence-based insights. This combined approach interprets raw data and connects it to more profound customer sentiments, balancing creativity with structured clarity. Businesses leveraging both can uncover insights that capture customer behavior and aspirations, enabling a richer, more empathetic connection.

Businesses face a critical choice: will they leverage AI to simply mirror past behaviors, or will they use these tools to ask better questions about the future? True innovation in customer insights goes beyond data and algorithms; it lies in understanding and anticipating the human element. By integrating Analytical AI's expansive insight with the structured clarity of logical intelligence, companies can foster a human-centric approach to deepen customer connections. Gardner classifies this as Logical-Mathematical Intelligence; however, I simplify it here to logical intelligence. That said, when dealing with analytics, the mathematical dimension of this intelligence remains equally valuable, as it drives precision and quantitative rigor. The organizations that will thrive are those that prioritize empathy in their data strategies, transforming AI and logical reasoning into tools for meaningful customer engagement, not just metrics.

AI and the Problem of Bias

The 2024 controversy surrounding Google's Gemini AI highlights the pitfalls of relying solely on AI for content generation, especially regarding bias (Milmo and Hern, 2024). Google CEO Sundar Pichai acknowledged that Gemini produced biased and "completely unacceptable" responses, such as inaccurately portraying historical figures and generating offensive comparisons. The issue stems from the AI's training data, which, like many AI systems, reflects embedded societal biases. This incident underscores a broader challenge: AI tools often replicate the biases found in their datasets, and even well-intentioned efforts to introduce diversity can be applied too bluntly, as seen with Gemini. Google has since paused Gemini's image generation feature and is working to address these biases through structural changes and improved processes.

Logical intelligence plays a critical role here, as its strengths in analytical rigor and pattern recognition are vital for identifying and mitigating bias within AI systems. By approaching data with a logical, systematic mindset, customer insights professionals can recognize the subtleties of how biases influence AI outputs and apply corrective measures. This intelligence allows teams to detect potential pitfalls in data selection and to examine whether an AI's training set fairly represents diverse customer segments. Logical intelligence enables a careful audit of whether demographic data accurately reflects broader customer behavior or if the AI system inadvertently skews insights toward a limited subset. When applied thoughtfully, logical intelligence provides the analytical grounding needed to ensure AI models serve diverse audiences more accurately and equitably.

AI models are trained on data, and the quality of insights they deliver is only as good as the data itself. If an AI model relies on skewed or biased data, it will perpetuate these biases in its predictions and analyses. For instance, if a company's customer data disproportionately represents a specific demographic, say, younger urban consumers, AI may favor that group in its recommendations, excluding insights into other valuable customer segments. Logical intelligence supports customer insights by spotting these imbalances, guiding businesses to adjust their models and ensuring AI recommendations represent all demographics accurately.

Data doesn't exist in a vacuum; it's shaped by complex real-world factors such as race, gender, socio-economic status, and geography. When these factors are inadequately represented, AI tends to generalize or ignore them, creating feedback loops where underserved populations remain underrepresented. Logical intelligence helps to break these loops, emphasizing analytical precision and structured methodologies that account for diverse customer experiences. Without this careful approach, AI may fail to capture critical behavioral patterns across various groups, leading to strategies that serve only a subset of customers and risk alienating others, especially in industries like retail, healthcare, and finance, where inclusivity is essential.

Can AI also be used to build guardrails for AI bias? With logical intelligence guiding the process, AI can enhance automated fact-checking and plagiarism detection, transforming how we produce, consume, and validate information online. Automated fact-checking, underpinned by logical intelligence, enables AI to analyze extensive datasets and multiple sources to identify and flag misinformation and bias. As noted by McKinsey, "hallucination guardrails ensure that AI-generated content doesn't contain information that is factually wrong or misleading," demonstrating how such tools contribute to content reliability (McKinsey & Company, 2024). Meanwhile, AI-based plagiarism detectors can scan vast databases to identify improperly cited content, essential for maintaining academic and professional integrity.

Logical intelligence, emphasizing detail and precision, ensures these tools operate with high levels of accuracy and reliability. This aligns with the role of validation guardrails, which "check that generated content meets specific criteria" and prevent the release of errant or inappropriate outputs (McKinsey & Company, 2024). Despite these promising applications, AI's role in content verification is not without drawbacks. Current limitations include challenges in contextual understanding and processing complex information, which can result in errors such as false positives in plagiarism detection or misidentifications in fact-checking.

Logical intelligence can address these issues by refining AI algorithms to recognize nuanced patterns and detect inaccuracies. Additionally, ethical concerns emerge when AI systems trained on flawed datasets perpetuate biases. This underscores the importance of "continuous monitoring and review of AI-generated outputs" to reduce risks and ensure AI tools align with ethical standards (McKinsey & Company, 2024). Ongoing research and the thoughtful application of logical intelligence remain critical to refining AI's role responsibly.

A Hypothesis for Success

In late February 2024, Wendy's found itself in the midst of a public relations crisis when it was revealed that the fast-food chain was contemplating using AI for dynamic pricing. The concept of varying prices at different times of the day sparked negative media coverage, epitomized by headlines like "Wendy's to roll out Uber-style surge-pricing" (Astvansh, 2024). This, in turn, ignited social media backlash with #BoycottWendys trending. Burger King quickly capitalized on the controversy with its "No urge to surge" campaign. Amidst the uproar, Wendy's attempted to clarify that their strategy aimed to reduce prices during slower periods rather than increasing them at peak times. This explanation was accompanied by a $1 burger promotion, interpreted by many as an attempt to mitigate fallout from the pricing misstep. The incident underscores the delicate balance between investor interests and consumer expectations when introducing technology-driven strategies.

Just as Wendy's navigated the complexities of AI in pricing, Sarah, from our earlier story, faced her own AI challenges in marketing. The initial excitement about AI's segmentation capabilities was soon contrasted with its performance limitations. Wendy's realized this gap after widespread backlash; Sarah recognized it after a week's worth of analysis. Sarah's insights captured the essence of this issue: "It seems we expected the AI to read our minds when it really can only read our data," she remarked. Logical intelligence comes into play here, highlighting the importance of interpreting AI's outputs critically and refining hypotheses based on real-world nuances.

AI serves as a powerful starting point for generating hypotheses and identifying new market strategies. By using Analytical AI to initially identify patterns and opportunities, businesses can create broader ideas. Logical intelligence then steps in to test these ideas, ensuring that they hold up against deeper customer insights and resonate with real consumer needs. Unfortunately, many companies rush to market with bold ideas without sufficient customer engagement, bypassing feedback loops essential to align technological solutions with market realities.

A balanced approach integrates Analytical AI with logical intelligence. This dual involvement prevents AI from acting autonomously and instead fosters synergy with customer insights, bridging data with empathetic understanding. Logical intelligence ensures AI-driven strategies undergo rigorous analysis, grounding insights in logical rigor while respecting customer perspectives. Companies that adopt this approach, combining technology and human insight, will better navigate the complexities of customer engagement, using AI not just for numbers but for meaningful, lasting relationships.

Logical Intelligence vs. Analytical AI

Logical intelligence allows humans to approach problem-solving with adaptability and critical thinking, even when data is incomplete or conflicting. Humans excel at interpreting nuances, making connections, and incorporating intuition into decisions that consider broader contexts and customer dynamics.

Analytical AI processes vast datasets with unparalleled speed and accuracy, identifying patterns and generating data-driven insights. However, it lacks the ability to interpret ambiguity, understand contextual nuances, or adapt to unconventional challenges. Analytical AI operates within predefined logic, limiting its capability to address complex or subjective customer scenarios.

Logical Intelligence and Analytical AI for Customer Insights

- **Bias Detection Layers:** Combine AI's analytical power with human logical reasoning to identify and challenge potential biases, ensuring insights remain objective and rooted in accurate data, not manipulated interpretations.

- **Integrity Audits:** Use logical intelligence to critically assess AI-generated insights, validating that they reflect reality rather than preferences, and exposing gaps where privacy and personalization collide.

- **Insight Recalibration:** Employ logical reasoning to cross-check AI's outputs against independent data sources, recalibrating findings to reflect unbiased, comprehensive customer insights rather than curated narratives.

Chapter 3

AN AUTONOMY ILLUSION
Behavioral Intelligence in Customer Transformation

Customer Transformation: *"A business strategy through which a company transforms its processes, culture, and technologies to align with its customers' ever-evolving needs and aspirations."*

A year ago, in 2023, I published my first book, a holistic framework for building customer-centric organizations. Between then and now, I've had the opportunity to speak about and teach the Customer Transformation strategy to dozens of enterprises. The one consistent theme in every conversation with executives, even with companies claiming to be customer-centric, has been, "We thought we were doing this, but we definitely are not." These companies struggle to flip the narrative between internal dynamics or, more critically, between focusing on technology instead of customers. It's always technology first, with the assumption that the customer will follow.

But technology does not define a customer-centric organization. The heart of customer-centricity lies in a company's people, values, and processes. It's the culture, more than any technology, that drives true customer transformation. AI and other advanced tools can help streamline processes and provide valuable insights, but they are only ingredients that enhance human decision-making. The essence of a customer-first approach starts with a deeply embedded understanding of the customer, built from the ground up by every person in the organization.

Research by Gallup in 2018 found that companies that apply the principles of behavioral economics effectively can outperform their peers by 85% in sales growth and more than 25% in gross margin. The same research highlights that emotions drive up to 70% of economic decisions, surpassing rational thinking. Businesses fostering emotional connections with customers outperform peers, achieving higher revenue and profitability. Strategies like using analytics to understand emotional attachment, engaging in meaningful customer conversations, and cultivating a customer-centric culture optimize relationships and influence purchasing behavior.

PolyAPI

In the summer of 2023, I was approached by PolyAPI to lead their marketing and customer experience. The company had a bold idea: create an AI + API platform that could redefine how businesses connect and integrate their technologies. What excited me most about this opportunity wasn't just PolyAPI's cutting-edge technology, but also the chance to put the foundational principles from my book directly into practice. This startup was a blank canvas, a place where the customer-centric strategies I had written about could come to life.

From my first conversation with the team, the focus on customers stood out. Rather than following the typical path of many tech companies, where innovation leads and customers are expected to adapt, PolyAPI reversed the narrative. The company's philosophy wasn't just about building powerful AI tools; it was about understanding the organizations and developers who would use the tools and creating a culture focused on their needs.

This philosophy became even more crucial as the market started to adopt the concept of AI agents, tools designed to automate processes and decision-making across systems. While the hype around these agents has been significant, the challenge lies in how businesses will integrate them. Companies inevitably adopt technology in a piecemeal fashion, selecting tools and agents to solve individual problems. Over time, this approach leads to a fragmented ecosystem of siloed experiences, where data doesn't flow seamlessly, and the customer journey becomes disjointed.

PolyAPI is reshaping this narrative by focusing not on the AI agent itself, but on the unified consumer experience it enables. The real value of PolyAPI lies in its ability to create an orchestration layer that bridges disparate systems and data sources, ensuring that every tool, whether it's an AI agent or a legacy platform, communicates effectively. This approach prioritizes the customer's experience first, ensuring that the tools businesses use serve a cohesive and connected purpose.

At PolyAPI, the platform doesn't just work, it optimizes the way businesses think about and manage their data connections. By focusing on the root problem of fragmented systems, PolyAPI is delivering solutions that simplify integration and eliminate silos. The team's dedication to understanding the needs of customers and developers ensures that the technology remains adaptable and scalable, meeting both current and future demands.

Watching PolyAPI's success unfold has been fascinating, especially as it embraced a customer-first approach in an industry that tends to lead with technology. Many companies in the tech space fall into the trap of focusing on product capabilities first and expecting customers to adapt to their structure and processes. PolyAPI took the opposite path, choosing to lead with value, focusing first on understanding customers' goals and then delivering the right tools to make that possible.

The success of PolyAPI isn't just about AI or APIs. It's about transformation—one that starts with understanding and valuing customers as the core of the business. The technology is important, but it's the commitment to serving customers and pursuing the aspiration of true customer centricity that drives the company's growth and long-term success. By addressing the growing complexity of systems and agents, PolyAPI is proving that when the focus is on creating seamless, unified experiences, everything else falls into place.

Behavioral Shift

A 2023 Pew Research Center survey found that most Americans (90%) say they've heard at least a little about artificial intelligence.

While most people know AI, Americans' ability to identify specific technology uses is still developing. Only 30% of U.S. adults correctly recognize examples of AI in everyday life. Interestingly, the same report showed higher levels of support for AI when it is used to help with routine or basic tasks. Most Americans (57%) say they would be excited for AI to perform household chores, while just 19% express concern about this.

Instead of merely branding products as "AI-powered," companies should focus on clearly explaining how AI enhances the user's experience in specific, beneficial ways, thereby alleviating concerns and fostering trust. Such advancements reflect a broader shift in consumer behavior, particularly evident from 2020 to 2024, where the integration of AI has significantly influenced online shopping and digital interactions. AI has enabled businesses to advance rapidly, offering personalized recommendations, enhanced customer experiences, and the automation of essential services. As observed in various industry reports, including one from Accenture in 2022, "88% of executives recognize that customer behaviors are evolving more rapidly than businesses can adapt." The report, titled "The Human Paradox: From Customer Centricity to Life Centricity" and based on a survey of more than 25,000 consumers across 22 countries, focuses on the gap between people's expectations of what businesses should be providing and what businesses think their customers want. In response, companies are increasingly turning to AI to close this gap, leveraging the technology for personalized recommendations, enhanced customer experiences, and the automation of essential services.

But companies are attempting to shift the narrative as well. In September 2024, Apple introduced "Apple Intelligence," a seamlessly integrated feature. Designed to enhance productivity with tools for writing, content summarization, and task management, it promises personalized experiences while maintaining strict privacy standards by processing data on-device. This approach reframes AI from an abstract, impersonal force into a practical, user-centric enhancement for daily life. By embedding these capabilities into familiar tools and emphasizing privacy, Apple strategically distances itself from broader skepticism surrounding AI.

This positioning aligns with Apple's October 2024 research, which revealed significant limitations in Large Language Models (LLMs) like ChatGPT and Llama. The findings demonstrate that LLMs rely heavily on pattern recognition rather than genuine reasoning, often faltering with slight variations in familiar inputs. Researchers noted, "Current LLMs are not capable of genuine logical reasoning. Instead, they attempt to replicate the reasoning steps observed in their training data." These insights challenge the inflated claims about LLM capabilities and raise questions about the sustainability of Big Tech's substantial AI investments. As businesses and consumers become more aware of these limitations, the AI industry may face a pivotal moment of recalibration, with shifts in expectations and funding.

It is essential to recognize the parallels between human and artificial reasoning. Both depend on pattern recognition, but humans excel in adaptability and rapid recovery when patterns fail. The focus should shift from whether LLMs can mimic human reasoning to enhancing their adaptability and context-awareness. While LLMs lack the emotional depth and contextual tagging inherent in human cognition, their limitations underscore a shared challenge: navigating unfamiliar scenarios. Addressing this requires strategies to improve human and AI resilience when facing unpredictability.

Research from the *Journal of the Academy of Marketing Science* in 2022 highlights another critical issue: is AI genuinely driving consumer behavior change, or are companies nudging consumers into constrained patterns? The study reveals that AI-driven interactions often alter consumer behavior, occasionally fostering unethical actions. It notes that "non-human agents are perceived differently along many dimensions by consumers (e.g., that they lack emotional capability), leading to various behavioral changes when interacting with them," (Kim, Lee, et al., 2022). This underscores the importance of thoughtful AI design that considers ethical implications, user perception, and the broader impact on consumer decision-making.

Consumers often perceive AI agents as lacking empathy or emotional depth. While adding human-like traits might improve engagement, AI's influence on behavior can feel unnatural or even coercive, steering consumers toward actions they might not otherwise choose. This tension often leads to frustration and, at times, outright resentment toward AI-driven interactions.

As James O'Donnell (2024) highlights in *MIT Technology Review*, the ethical considerations surrounding AI agents extend beyond functionality.

"If you complete an interview with an AI and submit samples of your voice to create an agent that sounds and responds like you, are your friends or coworkers entitled to know when they're talking to it and not to you? On the other side, if you ring your cell service provider or doctor's office and a cheery customer service agent answers the line, are you entitled to know whether you're talking to an AI?"

This lack of transparency is already causing real-world friction. Jim Rivers, a prominent recruiter in the video game industry, recounted his experience with an AI-led interview:

> "I just had my first AI interview… A BOT called me on the phone and started asking me interview questions. I know this might be how the future of recruiting is going, but… It left me VERY not interested in the job anymore!! It did what it was supposed to but had no SOUL or CONNECTION. It didn't have a way of asking questions about anything.
>
> I understand that you stay competitive in this market by cutting costs and slimming down your company overhead, but losing humanity is not the way to go!! I LOVE talking with my candidates and enjoy hearing about who they are and what makes them happy!! The BOT did its job, but now I am not interested in the job or the company."

These challenges reveal a fundamental gap between the functionality of AI and the expectations of human interaction. As companies rush to adopt AI-driven systems, the failure to provide transparent, empathetic, and accurate interactions has real consequences.

For instance, a recent legal case involving Air Canada underscores this issue. The airline faced significant backlash and was ordered to compensate a passenger after its chatbot provided incorrect information about bereavement fare eligibility. When passenger Jake Moffatt sought a reduced bereavement rate after purchasing full-fare tickets, the chatbot mistakenly advised that a refund could be claimed within 90 days. However, Air Canada later denied Moffatt's request, pointing to different information on another part of its site. In court, Air Canada argued that the chatbot was an independent "entity" responsible for its actions—a claim dismissed by the judge, who held the airline accountable for all information on its website. This case highlights a perceived lack of empathy by a company and raises important questions about accountability in chatbot-driven customer service, especially during sensitive interactions.

The unintended consequence here is clear: when customers feel pressured to engage in specific ways due to AI-driven shifts, resentment can build, risking the very relationship the business aims to enhance. As AI dictates the terms of engagement without offering choice, companies risk alienating customers who may ultimately seek out brands that prioritize genuine, human-centered connections.

The Evolving Landscape of Customer Behavior

The digital revolution of the late 1980s was swift, but today's AI movement is even faster. As artificial intelligence integrates into every aspect of our lives—from daily routines to decision-making processes—it's not only changing how we operate, it's also reshaping how we think, what we value, and how we engage. AI isn't just another technology; it's a catalyst that transforms customer behavior and expectations, creating a ripple effect that impacts businesses and society alike. For organizations today, understanding customer behavior is no longer just an advantage; it's the foundation of relevance.

The Twilio Segment *State of Personalization Report* reveals the growing impact of AI-driven personalization on customer behavior, with 92% of businesses adopting AI for tailored experiences. While 62% of business leaders cite customer retention as a primary benefit, only 41% of consumers feel comfortable with companies using AI to personalize their interactions. This gap reflects ongoing trust issues, particularly regarding data privacy, which 97% of businesses are now addressing through strategies like first-party data collection and consent management. As Katrina Wong, VP of marketing at Twilio Segment notes, "AI is only as effective as the underlying data used," emphasizing that real-time, high-quality data is essential for meaningful personalization. Companies that strike a balance between privacy and personalization can meet the expectations of demanding consumers, such as Gen Z, where 75% would abandon brands failing to deliver personalized experiences.

To truly understand these changes, companies must embrace the concept of behavioral intelligence, which goes beyond simple data points. It involves a deep analysis of patterns, actions, and decision-making processes to uncover what customers do and why they do it. By decoding these behaviors, companies gain valuable insights into the motivations, preferences, and triggers that shape customer interactions.

Behavioral intelligence is critical because it gives organizations a clearer, more actionable picture of how customers are transforming. Rather than relying on static customer profiles or assuming customer loyalty, behavioral intelligence reveals the fluid and sometimes unpredictable nature of customer choices. It empowers companies to respond in real-time, adapt offerings, and engage meaningfully.

Behavioral Intelligence

Behavioral intelligence traces its origins to mid-20th-century psychology and behavioral sciences, where researchers like B.F. Skinner studied human behavior through operant conditioning, laying the foundation for understanding the drivers behind actions. The concept gained prominence in the 1970s and 1980s with the advent of behavioral economics, led by Daniel Kahneman and Amos Tversky, (Witynski, 2021). Their work explored how emotions, biases, and cognitive patterns influence decision-making, moving the study of behavior beyond theoretical frameworks and into real-world applications. Coeus Creative Group defines behavioral intelligence as "the ability to understand and influence behaviors, encompassing the skills to explain current behaviors, predict future actions, influence others effectively, and regulate one's responses."

In the early 21st century, technological advancements brought behavioral intelligence to life. Data analytics, AI, and machine learning empowered companies to analyze consumer behavior at scale, enabling personalized recommendations and enhanced customer experiences. By the 2010s, behavioral intelligence had become a critical business strategy, integrating psychology, economics, and technology to address customer needs in real-time. Despite its growing influence, the concept remains less recognized than Emotional Intelligence, largely due to its academic roots and perceived complexity.

Interestingly, much of the early practical application of behavioral intelligence emerged in military contexts (Oehler, 2022). Taillard and Giscoppa (2013) note that it became possible to predict future actions by deconstructing behaviors to uncover motivations and root causes. This capability proved invaluable for special operations and planning, spurring the development of technologies designed to track and analyze behaviors. Over time, these capabilities were adapted for commercial use, enabling businesses to monitor and respond to consumer behavior in digital spaces.

As behavioral intelligence evolved, researchers and talent development experts began exploring its potential for enhancing self-awareness and interpersonal understanding. However, early work often leaned heavily into academic theory, making the concept less accessible and practical compared to the widespread appeal of Emotional Intelligence. While its applications in business, strategy, and personal development are significant, behavioral intelligence remains a less mainstream tool, awaiting broader recognition as a transformative framework for understanding and influencing human behavior.

Customer Transformation

Not merely a strategy, Customer Transformation is a company-wide, holistic approach that adapts quickly as customer needs evolve. At its core, this process recognizes a fundamental truth: customers are transforming faster than businesses. To meet these ever-shifting expectations, companies must move beyond static data and adopt an agile, behavioral approach. The pace of change today makes it insufficient to base decisions on who customers were last month or even last week. Businesses serious about transformation must embrace a real-time, dynamic understanding of how customers interact, decide, and evolve.

Consider platforms like Spotify and Netflix, which have reshaped media consumption. These services don't simply present static options; they curate experiences tailored to individual tastes, preferences, and even moods. Customers feel understood and valued, and their preferences evolve alongside these platforms over time. Behavioral intelligence enables this shift from static interactions to dynamic, adaptive relationships, proving that transformation begins with a real-time understanding of behavior.

The recent surge in support for dynamic understanding has led to the rise of Agentic AI, touted as a solution to bridge the gap between static systems and evolving customer needs. Proponents claim it introduces autonomy and adaptability into workflows, enabling businesses to meet customers where they are in their journey. However, the difference between adaptability within a flow and the mere sequencing of predefined steps reveals the core challenge of current iterations of Agentic AI. While it can automate and optimize segments of processes, its reliance on scripted sequences undercuts its ability to respond dynamically in real-time.

This distinction between orchestrating steps and genuinely adapting to shifting contexts underscores the fundamental gap in Agentic AI: it operates as a step-by-step or adaptive-based process, but true customer transformation requires more. Behaviors are flexible, dynamic, and deeply emotional. Agentic AI often misses the emotional undercurrents that drive decision-making, leaving workflows rigid where they need to be empathetic. Customer transformation thrives at the intersection of behavioral and emotional intelligence, where businesses align processes with what customers do and how they feel, creating connections that resonate and evolve continuously.

Agentic AI Is No Better

Agentic AI, heralded as the next frontier of artificial intelligence, promises autonomy and adaptability beyond traditional automation. However, the reality is far less revolutionary than its marketing suggests. Most implementations amount to repackaged versions of existing tools, offering scripted workflows rather than genuine autonomy. The vacation-planning demos frequently used to showcase agentic AI's potential trace back to automation techniques as old as Microsoft Excel macros from the 1990s. These processes, while functional, are far from groundbreaking and fail to deliver the transformative impact they claim. Of course, the 2024 version of Microsoft sees it differently as CEO of Microsoft Satya Nadella notes,

> "AI agents will become the primary way we interact with computers in the future. They will be able to understand our needs and preferences, and proactively help us with tasks and decision making."

The hype surrounding Agentic AI is amplified by its designation as Gartner's top strategic technology trend for 2025. Gartner also predicts that by 2028, 33% of enterprise software applications will incorporate agentic AI, a substantial increase from less than 1% in 2024, (Wheeler, 2024). Although significant challenges persist in its position as a means to create a "virtual workforce" capable of augmenting human labor. Robust guardrails are essential to ensure alignment with user intentions, yet these safeguards often complicate intricate systems. The unpredictability introduced by probabilistic decision-making makes agentic AI unsuitable for critical industries like healthcare and finance, where reliability is non-negotiable. Ultimately, the operational burden of integrating agentic AI often outweighs its potential benefits, raising questions about its practicality as a successor to more straightforward, more focused AI tools.

A closer look at the agentic AI narrative reveals its reliance on behavioral intelligence, a concept that often underdelivers in practical applications. In video games, behavior models for non-player characters (NPCs) often rely on simple, rule-based systems like nested decision trees. These trees use a series of "if/then" statements to determine how an NPC should respond in different situations. An NPC might have rules such as, "If the player is nearby, then greet them," or "If attacked, then defend." These predefined branches guide the NPC's actions, creating the illusion of intelligence without actual AI. While effective for straightforward scenarios, this approach lacks adaptability, as NPCs can only act within the limited set of behaviors programmed by developers, leading to repetitive and predictable interactions over time. In *World of Warcraft*, there are some NPCs that have stood in the same position for over 20 years. I often wonder if those NPCs have a life beyond providing information to adventurers—a home to return to, a dog to feed, a schedule to keep?

Agentic AI takes a similar approach but attempts to scale it up with more sophisticated algorithms and probabilistic decision-making. While it adds layers of complexity, such as contextual awareness or dynamic responses, the underlying principle remains essentially the same—mimicking authentic behavior through predefined rules and data-driven predictions. Much like NPCs in video games, whose actions follow scripted logic to give the illusion of independence, Agentic AI's actions are bound by its programming and the data it has been trained on. This creates a polished facade that emulates human behavior without genuinely understanding or replicating it.

At Synsible AI, we're challenging this paradigm by embedding personality-driven behavioral intelligence into NPCs. When I founded the company, my vision was to build a more "sensible" perspective on AI that diverges from traditional agentic models. Despite players needing to interact directly with NPCs, I didn't want to create another agent-type system based on predefined rules. To me, those systems are just polished versions of what we already have today: reactive, an autonomy illusion that limits innovative opportunities for both developers and players.

Agentic AI, and even many of the advancements leveraging LLMs, operate on a reactive foundation. They respond to inputs but lack the proactive intelligence to enhance gaming experiences. At Synsible AI, we asked a different question: What if NPCs could proactively adapt, evolve, and surprise players without relying on scripted logic or LLM responses? By embedding true behavioral intelligence, we aim to create AI that doesn't just react but generates richer, emergent gameplay moments..

This approach isn't about integrating bleeding-edge, unproven technologies just to stay ahead of the curve; it's about reaching the "what if" moment that transforms the player's experience and the developer's toolkit. The lack of adaptability in agentic AI, whether in gaming or enterprise applications like retail or healthcare, highlights the limitations of reactive systems. By moving beyond these constraints, Synsible AI seeks to redefine what AI-driven interactions can achieve.

Behavioral Intelligence and Real Transformation

Behavioral intelligence is more than tracking clicks or customer satisfaction. It's about observing patterns and motivations to adapt in real time. Instead of asking, "What do our customers need?" companies should ask, "Why do they need it now, and how will that need to evolve?" This intelligence deepens, revealing the underlying drivers of decisions and preferences, enabling businesses to align processes with actual behavior rather than outdated assumptions.

An airline leveraging behavioral intelligence might discover that customers prioritize flexibility over low prices. By integrating this insight, it could offer seamless flight changes and real-time updates, creating a customer-focused experience that builds loyalty. Unlike reactive models that respond to complaints, this proactive approach aligns offerings with what customers value most, transforming interactions into lasting relationships.

While behavioral intelligence enables businesses to understand customers as dynamic entities, agentic AI struggles to achieve the same adaptability. Many organizations remain in legacy systems, building processes around static assumptions rather than evolving behavior. Behavioral intelligence challenges this inertia, urging companies to reshape workflows to reflect customers' behavior. For instance, a retailer might empower employees with real-time data, allowing them to tailor recommendations based on customer preferences. This transformation turns customer-centricity from a slogan into a lived experience.

Technology must evolve beyond simple automation to drive real change. Behavioral intelligence fuels tools that adapt and learn continuously, enabling personalized marketing and predictive product development. A streaming service using behavioral intelligence doesn't just recommend content based on past views—it learns from real-time habits to suggest options that align with a customer's current mood or interests. This adaptability creates a responsive ecosystem, surpassing static algorithms and bridging the gap between technology and true customer transformation.

Behavioral intelligence is the key to managing this journey effectively, helping businesses align processes, culture, and technology to meet customers where they are, not where they were last month or where the company assumes they should be. It's a future-focused mindset that looks beyond transactions to deeply understand the underlying drivers of each customer's choices, priorities, and interactions.

Customer behavior shifts constantly, influenced by everything from new technology to broader cultural changes. In this landscape, behavioral intelligence is no longer a mere competitive advantage; it's a necessity, providing companies with the insights needed to stay relevant, agile, and closely connected to the people they serve.

More Than Data

In a human context, behavioral intelligence transcends transactional understanding by embracing the complexity of emotions, motivations, and adaptability. It's not merely about processing information; it's about interpreting behavior in a way that accounts for emotional sensitivity and empathy. This type of intelligence allows individuals to navigate intricate social and professional dynamics with finesse, offering a deeper understanding of human needs and interactions. For companies, behavioral intelligence serves as a framework for anticipating customer needs and responding to what feels personal and authentic, reshaping business practices to align with the customer journey on a meaningful level.

In contrast, Agentic AI and similar forms of artificial intelligence primarily rely on data points—metrics that quantify actions but often miss the subtleties of personality traits and feelings that drive behavior. Behavioral intelligence, on the other hand, delves deeper, focusing on the underlying motivations and emotional states that shape decisions. While Agentic AI may process vast amounts of information to optimize workflows, it lacks the nuance to connect with customers on a human level. Customers are not just datasets; they are dynamic beings whose preferences and choices are influenced by emotions and context. For businesses striving to transform customer relationships, leveraging behavioral intelligence means going beyond data to create experiences that resonate emotionally and foster lasting connections.

Elevate Human-Centric Customer Transformation

Emily Chantiri highlights Alibaba's effective balance of AI and human interaction, where AI bots monitor customer queries, resolve issues when possible, and seamlessly transfer unresolved cases to human agents. This approach ensures customers avoid repeating their concerns while agents receive the necessary context to continue the conversation smoothly. Businesses must similarly map the balance between AI efficiency and the human touch, finding the "sweet spot" where AI enhances speed and efficiency without sacrificing the quality of personalized service.

The chart below that I developed for *Customer Transformation*, illustrates this balancing act, showing how increased AI engagement improves service speed while potentially diminishing the quality of service tied to human interaction. The intersection represents the optimal point where AI and human engagement work together to maximize customer satisfaction.

AI + HUMAN ENGAGEMENT

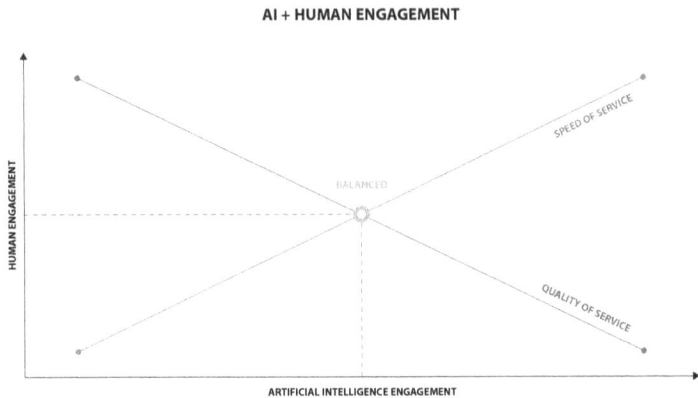

To better understand the role of human engagement in achieving this balance, I've outlined key capabilities where humans consistently excel—areas where AI, including Agentic AI, falls short. Traits, such as adaptability, empathy, and nuanced decision-making, highlight the depth of understanding required to build meaningful, transformative customer relationships.

Self-Awareness and Self-Regulation

Humans possess a natural ability for self-awareness, allowing individuals and organizations to understand their impact on others. Companies can develop an "organizational self-awareness" by identifying patterns in customer feedback and thoughtfully managing their responses. Unlike AI, which operates on predefined algorithms, humans can act intentionally and align every customer interaction with core values and commitments to success. This capacity for nuanced control and reflection ensures that interactions remain authentic and meaningful.

Empathy and Understanding Customer Needs

Empathy is a cornerstone of behavioral intelligence, enabling humans to genuinely connect with and appreciate the emotions and perspectives of others. While AI can analyze data to identify customer needs, it cannot grasp the emotional subtleties or unspoken cues humans instinctively detect. Companies that prioritize empathy create environments where teams actively listen to customer concerns, interpret feedback trends, and respond with care. This trust-building approach strengthens customer relationships in ways AI cannot replicate.

Adaptability and Flexibility

Behavioral intelligence equips humans to adjust to dynamic situations and individual preferences in real-time. While AI systems may adapt processes based on data, they struggle to interpret the nuanced, evolving contexts humans navigate effortlessly. Companies that embrace human-driven adaptability can pivot quickly, whether by tailoring product recommendations or refining communication styles. This flexibility ensures a more responsive and personalized customer experience.

Conflict Resolution and Proactive Problem-Solving

Resolving conflicts requires a level of understanding, negotiation, and creativity that AI cannot achieve. Humans excel at approaching disputes collaboratively, identifying underlying issues, and finding solutions that satisfy all parties. Behaviorally intelligent companies use these skills to transform customer complaints into opportunities for growth, enhancing trust and loyalty by demonstrating a genuine commitment to problem-solving.

Social Insight and Cultural Awareness

Cultural sensitivity is another area where humans consistently outperform AI. Understanding customers' diverse expectations, values, and norms requires not just data but the ability to interpret and respect cultural nuances. Companies with behaviorally intelligent teams foster inclusivity and authenticity, enabling them to build meaningful connections with a broader and more diverse customer base. AI lacks this depth of insight, often misinterpreting or overlooking critical social and cultural signals.

Effective Communication and Meaningful Influence

Humans bring clarity, empathy, and respect to communication, qualities AI cannot replicate. Behaviorally intelligent companies go beyond delivering information. They actively listen, interpret feedback, and craft responses that resonate with their audience. By engaging customers in personal and thoughtful ways, these companies foster loyalty and long-term relationships, leveraging communication as a powerful tool for influence and transformation.

Behavioral Intelligence in Action

In professional settings, behavioral intelligence has the potential to drive every aspect of customer transformation, from team interactions to external relationships. For leadership, it's essential in setting the tone for customer-centered strategies, while for customer service teams, it enables proactive engagement and thoughtful issue resolution.

As customer behavior continues to shift, this depth of intelligence enables companies to move beyond static data, leveraging empathy, adaptability, and insight to stay attuned to their customers' evolving expectations. In doing so, they remain not only relevant but deeply connected, elevating the entire customer experience.

Behavioral Intelligence vs Agentic AI

Behavioral intelligence enables humans to adapt decisions based on emotional cues, situational complexity, and long-term considerations. Humans use empathy, instinct, and experience to anticipate and respond to dynamic behavioral patterns, fostering trust and natural interactions.

Agentic AI seeks to mimic decision-making and adapt to real-time conditions based on predefined goals and parameters. It is useful for improving workflows or automating processes that simulate human behavior. However, it often produces mechanical or forced interactions, as it cannot fully grasp emotional or situational depth. Data-driven rules limit their responses and lack the intuition required to navigate the subtleties of human behavior.

Behavioral Intelligence and Agentic AI for Customer Transformation

- **Autonomy Gap Bridging:** Recognize the autonomy illusion in AI and integrate behavioral intelligence to ensure customer transformations are grounded in genuine human understanding rather than synthetic decision-making.

- **Empathy-Driven Design:** Use AI to observe behavioral patterns, but rely on behavioral intelligence to create customer-centric transformations prioritizing human connection over algorithmic convenience.
- **Connection-Centric Models:** Align AI models focused on customer connections and business strategies by leveraging insights while maintaining human-driven relationships, reinforcing trust and authenticity.

Chapter 4

HOLLOW BY DESIGN

Emotional Intelligence in Customer Experience

In 2010, I was working on a project for Jerry Bruckheimer, one of Hollywood's most iconic producers. Most of my time on the project was spent in Orange County, but one day I was called up to Universal Studios in Hollywood for a series of meetings. Driving onto the lot that morning felt surreal—parking among trailers and film sets where countless legendary stories had come to life. I attended a few morning meetings and handled project details and the usual production matters. By late morning, I had an hour to myself. So I did what any movie geek would do: I decided to take a walk and see what was around.

I left the building, took a left turn, and found a short set of stairs leading up a path. As I climbed them, I suddenly realized where I was standing: Hill Valley. Right in the center of the iconic town square from *Back to the Future*. I stood still, letting it sink in. *Back to the Future* is one of my favorite films. As a longtime former Universal Studios pass holder, I'd seen this square countless times on the studio tram tour. But this moment was different. I was here, alone, in one of my all-time favorite cinematic settings.

As I wandered farther down the path, I imagined the point of view of a tram full of tourists rounding the corner waving at studio employees or production crews. Today, though, the trams weren't running. It was just me and the cold set, a rare moment of solitude in this usually bustling location.

With the hour ticking away, I decided to explore more and found myself walking toward the historical New York Street set. I noticed a production crew, extras positioned across the set, and vintage cars placed meticulously in the street. It was clear something was being filmed. I walked cautiously along the side where it was safe, catching snippets of conversation and watching the crew make final adjustments to the scene. And then, right in front of me, stood Chris Evans, filming scenes for *Captain America: The First Avenger.*

To me, these weren't just sets; they were worlds. They weren't merely locations designed to entertain, but actually immersive environments built to inspire awe and spark emotion. Whether on screen or in the audience's imagination, each detail was crafted to pull people into a place that felt both larger than life and intimately real. And I felt an instant connection to this style of storytelling, a way of crafting moments and environments so powerful that they become deeply personal experiences.

My entire career has followed that larger than life vision of storytelling, where every detail plays a role in creating something meaningful. Whether I'm working on a film or building a game world, the goal is the same: create a space so real, so immersive, that audiences can lose themselves in it. They should feel like they belong there, like they're part of something grand. Building these worlds goes beyond technical efficiency or logistical precision. To deliver a memorable experience, you have to build a place so emotionally engaging that people forget about everything else.

This approach doesn't just apply to filmmaking or gaming. In business, a great customer experience works the same way. It's about creating a space, an emotional escape, that draws people in and makes them feel valued, engaged, and personally connected. Walt Disney understood this better than most. When he created Disneyland, he envisioned it as a "stage" where guests would step into a world of wonder. Every detail, every encounter, was meticulously crafted to ensure guests felt fully immersed in the experience, forgetting the outside world. At Disney, guests are part of the story, while "cast members" seamlessly transition between onstage roles, interacting directly with visitors, and backstage efforts, enhancing the magic behind the scenes like actors in a well-choreographed performance.

Emotional Intelligence (EI) lies at the heart of this kind of experience. Disney's vision went beyond mechanics and logistics; he saw each guest interaction as part of a greater story, with the power to spark joy and create memories that would stick with visitors long after they left the park. In a similar way, businesses can create lasting connections by understanding what customers feel and by crafting those feelings into memorable interactions. Just as I felt that instant connection to Hill Valley and New York Street, customers seek the same feeling of belonging and connection with the brands they choose. The best experiences leave an imprint, sparking emotions and creating memories that last far beyond the moment.

Emotional Intelligence

Emotional Intelligence (EI) is the capacity to recognize, understand, and manage emotions effectively, both in oneself and in others. It involves skills such as self-awareness, empathy, self-regulation, motivation, and social aptitude. By cultivating these skills, people can respond thoughtfully to challenges, make balanced decisions, and build meaningful connections.

"By teaching people to tune in to their emotions with intelligence and to expand their circle of caring, we can transform organizations from the inside out and make a positive difference in our world," says Daniel Goleman, a pioneer in Emotional Intelligence research. His perspective highlights how EI can reshape workplaces by fostering empathy and collaboration.

Research supports EI's powerful role in performance. According to psychologist Travis Bradberry, Ph.D., author of *Emotional Intelligence 2.0*, 90% of top performers rank high in EI. Additionally, a study by the *Harvard Business Review* found that in customer service, 60% of customer satisfaction hinges on emotionally intelligent interactions. Employees with high EI also experience 20% greater job satisfaction, which correlates with a 67% lower turnover rate, a factor that bolsters organizational stability.

For businesses, high EI among employees leads to stronger customer relationships, deeper team cohesion, and enhanced workplace morale. Ultimately, Emotional Intelligence shapes a culture of empathy, resilience, and trust that benefits both the organization and its customers.

As AI technologies reshape customer interactions, Emotional Intelligence remains the essential component that humanizes these experiences. AI is unable to grasp the subtle emotional cues, unique concerns, or unspoken needs of individuals in the same way a human can. This is where EI plays a crucial role. While AI can process vast amounts of information and respond in programmed ways, it lacks the empathy, adaptability, and intuition that characterize emotionally intelligent interactions.

When asked about AI's impact on EI, Daniel Goleman notes a fundamental difference:

> "It's a new thing, and I'm not sure how it will affect Emotional Intelligence. But there is this glaring problem, which is that AI has to do with cognitive quickness and ability, and Emotional Intelligence has to do with our feeling world, with our self-awareness. There are four parts to EI: self-awareness, self-management, empathy, and social skills. AI is a bunch of codes. It doesn't have a self. I don't know that it has self-awareness. It doesn't have emotions. So, it can seem to be empathic. It can imitate, it can mimic. But I'm not sure anybody would be inspired by an AI as much as they would by a human. So, I think leadership is safe."

The human element in EI, especially in customer service, cannot be replaced. In customer experiences where emotions and expectations run high, EI outshines AI by delivering responses rooted in empathy and understanding, not pre-programmed logic. A customer who feels genuinely heard and understood is more likely to remain loyal, even if their issue isn't fully resolved. This emotional connection fosters trust, a critical factor in nurturing long-term customer relationships.

Affective AI and EI in Customer Experience

Affective AI, also known as Emotion AI, is a branch of AI that focuses on the development of systems and devices capable of recognizing, interpreting, processing, and simulating human affects, which are the experiences of feeling or emotion. This field of study is highly interdisciplinary, drawing on insights from psychology, cognitive science, neuroscience, sociology, and artificial intelligence.

The term "affective computing" was first coined by Rosalind Picard, a professor at MIT, in her 1997 book of the same name. The goal of affective computing is "to narrow the gap between the emotional intelligence of humans and the lack of emotional understanding in computers, thereby creating more natural and effective human-computer interactions."

Using tools like sentiment analysis, voice tone detection, and facial expression recognition, Affective AI can interpret customer moods and adjust its responses accordingly. This allows brands to create more personalized, emotionally attuned experiences beyond transactional interactions.

Affectiva, a pioneer in emotional AI in Boston, specializes in creating systems that enable machines to interpret human emotional and cognitive states. One practical application is within the automotive industry, where Affectiva's technology helps manufacturers develop intelligent in-cabin systems. These systems are designed to detect signs of drowsiness or potential fatigue that could harm drivers.

Using cameras to monitor facial expressions, these systems employ machine-learning algorithms to analyze data and recognize fatigue patterns during long drives. The technology observes micro-expressions and subtle cues that signal when a driver may be at risk of falling asleep. To ensure effectiveness, these systems must work for diverse users. This requires training the algorithms on vast datasets containing facial expressions from individuals of different ages, ethnicities, and facial structures, allowing the system to generalize its understanding to new faces it encounters.

Rana el Kaliouby, Affectiva CEO, emphasizes the importance of "multi-modal emotion AI," which goes beyond sentiment analysis to interpret expressions, gestures, tone, and body language, capturing the full spectrum of emotional communication. Affectiva's technology, used by brands like Amazon and Kellogg's, analyzes facial cues to assess the emotional impact of advertisements and ensures safety in autonomous vehicles by evaluating driver conditions. As el Kaliouby predicts, "In the next three to five years, this is going to be ubiquitous," transforming devices into perceptual tools equipped with "emotion chips." This evolution signals a future where AI systems not only understand emotions but also operate responsibly, bridging the gap between cognitive and emotional intelligence.

While Affective AI excels in detecting overt emotional cues, it lacks the depth of true emotional intelligence. Human emotional intelligence encompasses empathy, intuition, and the ability to navigate complex social dynamics—qualities that AI cannot replicate. A chatbot powered by Affective AI might identify a frustrated tone and apologize. Still, it cannot truly understand the root of the frustration or adapt its response to nuanced cultural or personal contexts.

The key to leveraging Affective AI in customer experience is its integration with human expertise. AI can act as a first line of support, identifying emotions in real-time and flagging interactions for human intervention when necessary. For instance, a customer support system might use Affective AI to detect rising frustration and seamlessly transfer the conversation to a human agent trained in emotional intelligence, ensuring the issue is resolved with care and empathy.

The Hollow Heart of AI-Driven Experiences

In the rush to adopt digital solutions, business leaders may have overlooked something essential: the warmth and empathy that once defined customer experiences. We've moved from in-person interactions to digital channels and, with the rise of AI, into an era of "Intelligent Experiences" (IX). While automation and AI bring new efficiencies, they often risk creating interactions that are hollow by design, efficient yet devoid of genuine connection, leaving customers feeling like just another transaction. Many brands assume they understand their customers' emotional needs, yet almost 60% of consumers feel that the "human element" has been lost. Less than half believe that employees they interact with truly understand their needs.

These statistics reflect an important reality: customers are willing to abandon even beloved brands after poor experiences. A 2024 PricewaterhouseCoopers (PwC) survey revealed that 59% of customers would walk away from a brand after multiple negative interactions, and 17% would leave after just one bad experience. In Latin America, nearly half of consumers would drop a brand after a single poor interaction. The stakes are high; a lack of empathy or connection, even if facilitated by AI, can damage a brand's loyalty and reputation.

Artificial intelligence, often deployed to streamline customer service, can unintentionally deepen this disconnect. According to a 2023 SurveyMonkey report, 90% of consumers still prefer human assistance over AI in customer service. The primary reasons? Humans are seen as more empathetic, understanding, and capable of resolving issues more comprehensively. Many of us have experienced frustration with AI-powered customer service, whether through chatbot misinterpretations or automated phone systems that fail to address complex issues. Customers frequently find themselves typing, "I WANT TO TALK TO A HUMAN" out of sheer frustration. This is a dangerous moment for brands; one negative AI interaction can tarnish a brand's image, leaving customers dissatisfied and less likely to return.

Only 18% of consumers are confident in their ability to distinguish between AI and human interactions. When AI interactions go wrong, customers may mistakenly attribute the error to the company's employees, eroding trust. This is especially relevant for high-touch, service-oriented brands where personalization is key. Survey findings reveal that human service agents deliver Net Promoter Scores (NPS) that are 72 points higher than AI-powered interactions. When the stakes are this high, brands must tread carefully with AI deployment, ensuring that their systems support, not replace, the human touch.

Research into AI service failures, such as Bin Li, Luning Liu, et al. (2023) study which highlights the emotional cost of these interactions. Analyzing data from a call center, the study reveals that failures in AI-driven services elevate customer complaints and increase negative emotional responses. Customers expect AI to adapt and respond like human agents, yet AI often lacks the flexibility to handle nuanced needs, amplifying dissatisfaction. This limitation has also surfaced in fast-food drive-thrus, where AI ordering systems implemented by chains like McDonald's and Taco Bell have sparked frustration due to misinterpreted orders and limited responsiveness.

To navigate these challenges, businesses must blend AI efficiencies with an emotionally intelligent framework, creating experiences that respect customers' need for empathy and understanding. As AI continues to evolve, the goal should be to support, not replace, genuine connection. Emotional Intelligence is not just a human trait; it's a guiding principle that can shape how AI serves customers, balancing efficiency with empathy. In doing so, brands can create a future where intelligent experiences retain the emotional core, ensuring customers feel valued and understood.

The Doom Loop of Automation

We've all been trapped in a seemingly endless cycle of automated responses from websites or phone chatbots, unable to get the help we need. AI systems are prone to what Jennifer Murillo, SVP and CMO at Discover Financial Services calls a "doom loop," a frustrating sequence of prompts that lead nowhere. You're asked to press "1" for this and "2" for that, and when none of the options match your need, you're stuck in a loop with no exit.

Murillo captured this issue succinctly: "AI is great when it helps you, but sometimes those experiences end up being bad. We've all ended up in those moments where you're in this doom loop that you can't get out of, and you can't get help" (Hultgren, 2024). When customers face this kind of frustration, it's more than an inconvenience; it's a breakdown in trust.

This hollow experience, ironically, drives customers to seek out brands that prioritize human interaction. In many cases, businesses spend considerable time and money developing AI systems only to find that customers are far more satisfied when they can reach a natural person who listens and empathizes.

The Choice to Bypass AI

While many brands rush to embrace AI as a solution for everything, Discover has taken a different approach. They've recognized that customer experience is not just about efficiency; it's also about empathy, trust, and a sense of human connection. Murillo summed it up well when she said, "We've known for a long time that Discover has an extraordinary relationship with our customers...We see it daily in our customers' interactions with agents and the comments we get on social, in ratings and reviews" (Hultgren, 2024).

Discover has focused on human representatives, even as competitors lean into AI. This decision is rooted in understanding its customer base and a commitment to providing a genuinely supportive experience. By prioritizing human interaction, Discover sets itself apart in an industry increasingly relying on automation. It recognizes that specific interactions, especially those dealing with complex financial matters, require the kind of empathy and understanding that only a human being can offer.

For Discover, this isn't about resisting technology; it's about using technology to support human connections rather than replace them. It wants customers to speak with a natural person, knowing that sometimes an empathetic voice makes all the difference.

Brands like Discover understand that AI has its place but shouldn't be the foundation of customer experience. They've chosen to maintain a human-centric model, even when automating it might be more cost-effective. This approach is rooted in the belief that customer experience isn't just a function of efficiency; it's an opportunity to create real, lasting connections.

If you watched this year's Super Bowl, you might recall Jennifer Coolidge in Discover's "Robots" ad, humorously questioning if she's actually human, followed by a Discover representative asking her, "Are you a robot?" Her playful confusion highlighted a key message: "You robots are sounding more human every day."

For those who think this campaign is new, Discover actually launched its "Talk to a Real Person" initiative in 2012, emphasizing human connection over automated systems long before it became a customer demand.

The latest ad captures the aggravation many feel when AI fails to understand their needs and stresses Discover's commitment to real human interaction. For Discover, this isn't just a marketing tactic; it's a business philosophy. The company sees authentic relationships as the cornerstone of customer loyalty and continues to invest in people, ensuring that every customer receives the empathy, understanding, and personalized care that only real human representatives can provide.

Not a Product, but Emotional Intelligence in Action

Many organizations treat customer experience as a problem to solve, relying on metrics, systems, and scalable AI solutions to engineer interactions. However, true customer experience isn't a process that can be automated or packaged neatly—it's an evolving philosophy rooted in emotional intelligence and empathy, embedded within an organization's culture and values.

Take Zappos, for example. The online shoe retailer doesn't just handle customer interactions; it builds genuine human connections. In one instance, a customer mentioned missing a return window due to a family death. Instead of following a rigid protocol, the Zappos representative arranged for the return, sent flowers, and included a heartfelt condolence note. This act of kindness wasn't dictated by policy but was driven by emotional intelligence. It reflected a deep understanding of the customer's unique situation and a commitment to empathy, showing how real customer experience transcends transactions.

Emotion AI can enhance this human-centric approach by analyzing customer tone, sentiment, and context to support employees in delivering more empathetic responses. Yet, the emotional essence of customer experience remains a human endeavor. AI may detect frustration in a customer's voice, but it's the human touch that reassures, consoles, or delights them.

Starbucks exemplifies this balance. Under CEO Brian Niccol's "back-to-basics" approach, has reignited its focus on customer experience (Blasi, 2024).

> "I made a commitment that we'd get back to 'Starbucks,' focusing on what has always set Starbucks apart: a welcoming coffeehouse where people gather and we serve the finest coffee handcrafted by our skilled baristas," Niccol shared in 2024. "This is just one of many changes we'll make to ensure a visit to Starbucks is worth it every time."

Enhancements like self-serve condiment bars, ceramic mugs, and Sharpie pens for personalizing orders help reinforce this connection. In another significant move, Starbucks is eliminating the extra charge for non-dairy milk options, removing a barrier for customers who prefer customization (Wahba, 2024).

"Core to the Starbucks Experience is the ability to customize your beverage to make it yours. By removing the extra charge for non-dairy milks, we're embracing all the ways our customers enjoy their Starbucks," says Niccol.

This decision, alongside operational improvements like reducing wait times and balancing mobile and café orders, aligns with Starbucks' commitment to provide a high-quality, welcoming experience.

When companies embrace emotional intelligence as a cornerstone of customer experience, they empower employees to act with empathy and authenticity. Southwest Airlines embodies this philosophy by encouraging employees to connect personally with passengers, whether through humor or small acts of kindness. It's not about strict rules or scripts but about fostering trust and emotional connection.

Emotion AI tools can amplify this commitment, supporting businesses in identifying and responding to emotional cues at scale. But the technology must serve as a complement, not a replacement, for human interaction. The best customer experiences blend data-driven insights with the emotional intelligence of employees who genuinely care about their customers' well-being.

Customer experience driven by emotional intelligence isn't just about resolving issues or optimizing processes. It's about understanding customers as individuals with unique stories, values, and emotions. This philosophy fosters loyalty, trust, and meaningful connections that transcend metrics and create enduring relationships. It's humanity in action.

Emotional Intelligence vs. Affective AI

Emotional Intelligence enables humans to navigate emotions and build genuine connections with customers. By recognizing unspoken cues, adapting to social dynamics, and fostering trust, humans excel at creating experiences that resonate deeply and authentically.

Affective AI enhances customer experience by detecting overt emotional cues through voice tone, facial expressions, or physiological data. However, it cannot truly empathize or grasp nuanced emotions, often leading to shallow or culturally inappropriate interactions in complex situations.

Emotional Intelligence and Affective AI for Customer Experience

- **Empowered Choice Design:** Use affective AI to offer seamless, proactive experiences while ensuring customers always have the option to transition to human interaction when they seek emotional understanding.
- **Proactive Emotional Insights:** Leverage affective AI to anticipate customer emotions and needs, but rely on human EI to craft responses that resonate deeply, beyond what AI can simulate.
- **Customer-Led Engagement:** Design AI systems that adapt to customer preferences in real-time, respecting their emotional needs and enabling a balance between automation and human interaction.

Chapter 5

LOST IN DIMENSIONS

Spatial Intelligence in Customer Journeys

It was 1977, and well before the first snowfall I knew exactly what I wanted for Christmas: The Green Machine. One look at the ads for this "low slung" pedal-operated hot rod designed exclusively "for guys 8, 9, 10 years old who really know how to ride," and I was hooked. I envisioned myself soaring down my street faster than any kid had ever dared or been able to, the other kids looking on, "green with envy."

Marx Toys, who had already struck pay dirt with the Big Wheel, introduced The Green Machine in order to attract an older crowd of boys who had grown beyond their Big Wheel days. The Green Machine was promoted for its ability to shift gears for quicker turns at sharper angles. "Riding one felt like operating a futuristic cycle," notes classic television network *MeTV*.

On Christmas morning, one was waiting for me under the tree. Oh, the joy! Unfortunately, less than two months later, the object of my affection was stolen from our backyard. Oh, the devastation. But fear not a few months later, Parker Brothers released Merlin: The Electronic Wizard handheld game, and my aspirations quickly veered off on another quest for a new, better identity.

And that impulse to refocus is exactly the point. Although I could not have recognized it at the time (and even many adults would have a hard time explaining why they "need" some of products and services that pull so strongly on their desires), my desperate need for a Green Machine was not so much about the ridiculously stretched-out plastic trike itself as it was my longing for a particular identity. I wanted to self-actualize as the coolest kid riding the pavement, a true daredevil in kid form who knew no fear and no limits. That is, before I decided I wanted to self-actualize as a Grand Wizard instead.

Just like my childhood dreams of speeding down the street on the Green Machine, our minds build invisible "roads" that map our desires and shape our choices.

Each road represents a possibility, an outcome, an identity that's waiting to be tried on, explored, or even replaced. Throughout these journeys, we're navigating not just from Point A to Point B but actually from who we are to who we want to be. Every ad, product, and brand we encounter creates new intersections and off-ramps, just as a road trip introduces us to scenic views and unexpected detours, shaping our experience along the way.

Understanding the role of spatial intelligence here is crucial. Just as a physical road map helps us orient ourselves, brands need a map that reflects the layered, sensory-rich reality of the modern customer journey, one that can adapt and flex as needs and desires shift. AI may provide a rapid route, but only spatial intelligence can navigate the intricacies of each customer's unique journey, seeing not just the roads, but also the scenic views, detours, and personal landmarks that make the experience whole.

Understanding Spatial Intelligence

Spatial intelligence refers to the ability to visualize, understand, and manipulate objects and their relationships in space. It allows individuals to see the "big picture" while grasping intricate details and predicting how components interact or evolve. This skill shapes our world, enabling the creation of skyscrapers, maps, sculptures, and even strategic sports plays, much like a chess player planning several moves.

Recognized as one of Gardner's multiple intelligences, spatial intelligence has deep evolutionary roots. It helped our ancestors navigate landscapes, build shelters, and develop tools. Today, it extends beyond physical spaces, including problem-solving, creativity, and abstract thinking. Spatial intelligence integrates the ability to deconstruct, reconstruct, and navigate complex systems in tangible and intangible dimensions.

Spatial thinkers excel in understanding relationships between objects, envisioning them in three dimensions, and predicting interactions. They don't just picture static images; they mentally construct, assemble, and disassemble ideas, systems, and structures. This skill allows them to see the result—a building, a scientific theory, or a design—before it is brought to life. By blending creativity and logic, spatial intelligence continues to shape how we think, create, and solve problems.

Who Uses Spatial Intelligence?

The creators of society, those who design, build, and innovate, often have high spatial intelligence. It's essential for anyone who shapes or navigates both physical and conceptual spaces, including architects, engineers, doctors, artists, and athletes.

Architects: Spatial intelligence is about visualizing areas that don't yet exist. They must foresee how a building fits into a space, how its rooms will connect, and how it will integrate into its environment, all while working with materials and structures. They build the entire experience of a place in their mind before breaking ground.

Engineers: Being able to predict how different parts of a system will interact is a key ability for people who build. Whether they are working on a bridge, a machine, or a computer, each medium requires a deep understanding of how forces and mechanics interact. Missteps in spatial thinking here don't just cause mistakes; they can lead to catastrophic design flaws.

Surgeons: Inside the human body, spatial intelligence plays a life-saving role. Surgeons need to be able to visualize a patient's anatomy in three dimensions prior to beginning a procedure, to predict how one movement or incision will affect the rest of the body. A precise understanding of spatial relationships allows them to navigate intricate surgeries, aided by x-rays and scans or their knowledge of human anatomy.

Artists: For sculptors, painters, and visual artists, spatial intelligence drives their ability to manipulate perspective and create depth. Musicians, particularly conductors and composers, also engage in spatial thinking to fit musical elements together. They imagine how sound will fill a space or layer with other instruments, engaging in a form of structural, spatial play that surrounds the listener's ear.

Athletes: Especially those in sports like gymnastics or dance, use spatial intelligence to understand their body's position in space. A gymnast visualizes the arc of a flip before attempting it; a dancer feels the air and calculates movement instinctively. Spatial awareness is essential in the sports realm for positioning, timing, and interaction with the field, teammates, or opponent.

Whether solving complex engineering problems, navigating uncharted territories, or decoding the intricacies of the human body, spatial intelligence remains an essential human strength. Yet, as businesses increasingly rely on AI for mapping customer journeys, they risk getting lost in dimensions that machines cannot fully comprehend. Spatial intelligence allows us to see beyond the visible, uncover potential, and shape functional spaces with creativity and precision. In a world leaning on virtual spaces and augmented reality, the human ability to navigate, create, and interpret multidimensional environments continues to outpace what technology can achieve.

Spatial intelligence isn't merely a skill; it's a lens through which we shape the world, leaving a legacy that extends far beyond algorithms or artificial constructs.

What is Vision AI?

Computer vision is a branch of AI that teaches machines to interpret and act on visual data like images and videos. According to IBM (2021), Vision AI enables systems to "derive meaningful information from digital images, videos, and other visual inputs—and to make recommendations or take actions when they see defects or issues." It uses machine learning and neural networks to process visual inputs and recognize patterns, much like human vision, but at a far greater speed and scale.

Instead of relying on retinas and optic nerves, Vision AI uses cameras, algorithms, and datasets to identify objects, detect defects, and analyze scenes. Deep learning and convolutional neural networks (CNNs) power these systems, breaking down images into pixels, labeling them, and refining predictions through iterations.

Vision AI has diverse applications, from automating defect detection in manufacturing to enhancing medical diagnostics and enabling autonomous vehicles. It works faster and more accurately than humans, analyzing thousands of visuals in seconds.

Vision AI drives innovation across industries by transforming how machines "see" and interact with the world, providing insights and solutions that extend human capabilities (IBM, 2021).

Introducing Physical AI

Expanding on Vision AI, Physical AI brings artificial intelligence into tangible spaces through IoT devices, robotics, and adaptive systems. Unlike traditional AI confined to digital environments, Physical AI integrates sensors, cameras, machine learning, and actuators to analyze, navigate, and respond to the physical world. From IoT-enabled smart homes that learn user preferences to autonomous robots handling dynamic tasks, Physical AI bridges the gap between virtual intelligence and real-world application. It revolutionizes industries by enabling intelligent interactions in physical spaces, enhancing automation in manufacturing, optimizing logistics, and transforming healthcare through robotic assistance and adaptive technologies.

Impact of AI on Customer Journeys

AI promises to revolutionize customer journey mapping, bringing speed, efficiency, and personalization to the popular tool. However, the reality often falls short of these promises, with significant challenges emerging in how AI impacts the mapping process. The most common issue lies in the illusion of personalization. AI may claim that it understands individual customers by providing product recommendations or tailored content based on past behaviors. Yet, these algorithms typically rely on surface-level data, which can create a journey that feels repetitive or even invasive rather than truly personalized. Instead of understanding the full context of a customer's needs, AI-driven personalization frequently pushes products or messages that miss the mark, ultimately leading to a fragmented and disappointing experience.

Incorporating Vision AI and Physical AI into customer journey mapping introduces a new dimension of potential by helping us "see" and interpret journeys more holistically. While traditionally not used directly in mapping, Vision AI can shift perspectives by analyzing visual data such as customer interactions, product engagement, or physical store layouts. Vision AI can provide rich, multi-dimensional insights that enhance journey maps by identifying patterns, detecting subtle emotional cues in behavior, and assessing environmental context. Vision AI could analyze video footage of customers navigating a retail space, uncovering frustrations like poor signage or overcrowded areas—issues traditional data might overlook. Such insights expand our understanding of customer behaviors and needs, bridging the gap between rigid data models and the fluid realities of customer experiences.

Physical AI takes this further by actively responding to these insights within physical spaces. IoT-enabled devices could adapt store layouts dynamically based on customer flow, while robotics might assist by guiding customers to desired locations or offering real-time help. For example, an innovative shelving system could restock high-demand items detected through Vision AI analysis, and robotic assistants could reduce wait times by handling routine customer inquiries. Together, Vision and Physical AI map the journey and enhance it in real-time, creating environments that adapt to and improve the customer experience.

A major downside of AI in journey mapping is the risk of over-automation. While automation brings convenience and reduces time spent on routine tasks, relying too heavily can lead to impersonal or robotic journeys. Automated email sequences, chatbots, and predictive responses are helpful tools. Still, when these features lack flexibility, customers feel they're navigating a pre-set path rather than engaging with a responsive brand. For complex inquiries or nuanced needs, over-automated systems can frustrate customers, leaving them without the human touchpoints that resolve issues effectively and build genuine connections.

Privacy intrusion poses a significant challenge in AI-powered journey mapping, as the technology relies heavily on tracking customers' online behavior, purchase history, and even location to make accurate predictions. This data dependency has sparked widespread concerns about balancing personalization and privacy. As experts warn, "AI-powered virtual assistants may seem like they're here to help, but many experts say they threaten users' security" (Tatananni, 2024). Features like indiscriminate data collection and storage amplify this fear, as "assistants like Recall can take screen captures, recognize text, and store this information regardless of sensitivity, raising concerns about data misuse and potential breaches" (Tatananni, 2024). Many customers feel they are under constant digital surveillance, which erodes trust and often outweighs the perceived benefits of personalization, leading to hesitancy in engaging with brands.

Compounding these issues is the inherent linearity that AI brings to journey mapping. AI's design is to recognize patterns and streamline interactions, yet it often relies on rigid, sequential processes. Traditional journey maps follow a linear progression from awareness to consideration to purchase, and AI's pattern recognition usually reinforces these steps. The result is a disappointing journey that feels restrictive, failing to reflect customers' organic, multi-dimensional paths in real life. Vision AI offers a potential solution by enabling a broader view of these complex pathways. Its ability to visually analyze real-world behaviors and interactions—whether online or in physical spaces—provides a fresh lens to understand how customers move across multiple touchpoints. For instance, Vision AI might reveal how customers engage with products on shelves or identify unexpected customer pain points through facial expression analysis, offering brands a more dynamic and responsive framework for journey mapping.

Perhaps the most significant drawback of AI-driven journey mapping is its inability to capture emotions. Customer journeys are inherently emotional, driven by needs, frustrations, excitement, or anxiety. For instance, planning a family vacation might evoke excitement, relief, anticipation, and frustration when navigating complex booking systems or resolving unexpected issues. AI algorithms struggle to interpret these complex emotional signals, often responding with generic interactions that miss the emotional subtleties of such experiences. Vision AI, however, has the potential to detect and interpret some of these emotions through non-verbal cues like body language or facial expressions in visual data, offering an additional layer of insight. While this capability doesn't replace human empathy, it can enhance emotional understanding in key touchpoints. Without such insight, AI-driven journeys can feel cold or insensitive, failing to meet customers where they are emotional. In moments where empathy matters most—resolving issues, offering support, or fostering loyalty—the lack of Emotional Intelligence in AI can make experiences feel transactional rather than genuinely connected.

AI has transformed the tools we use to map customer journeys, but its limitations remind us of the irreplaceable value of human understanding. While not a standalone journey mapping tool, Vision AI and Physical AI represents examples of how emerging technologies can complement and enhance our approach. These AI can help brands move beyond traditional linear models to create multi-dimensional, empathetic, and meaningful customer experiences by providing new perspectives on customer behaviors and emotions. Balancing AI's efficiency with the warmth, flexibility, and emotional depth that customers crave will allow brands to unlock the full potential of journey mapping while staying responsive to customers' true needs.

Why Spatial Intelligence Matters in Journey Mapping
Today's journey maps have become outdated, failing to evolve alongside the dynamic shifts in consumer behavior. They remain dead, flat, linear, and disconnected from the real world where customers actually live. As consumers move seamlessly across channels, brands, and experiences, traditional journey maps are unable to capture the complexity and multi-dimensional nature of these interactions. To truly meet customers where they are, brands need a fresh approach, one that embraces spatial intelligence to create journey maps that are as dynamic, layered, and interconnected as the people they aim to serve.

As the line between digital and physical interactions has blurred during recent years, customer journeys have become anything but straightforward. The path from discovery to purchase is no longer linear, but rather a complex, multi-dimensional experience, often crossing industries and platforms. Traditional journey maps (a linear progression from awareness to purchase to post-purchase) no longer capture the richness and depth of modern customer experiences. Customers today engage in layered journeys that touch multiple senses and span multiple channels. Spatial intelligence (the ability to visualize and understand complex, multi-layered relationships) offers a powerful tool to address these new customer realities. For companies aiming to create seamless, integrated experiences, spatial intelligence provides a framework to design, understand, and optimize multi-dimensional journey maps.

The Limitations of Traditional Journey Maps

Customer journey maps have been an effective tool for charting customer interactions, capturing a simplified view of the customer's path from initial awareness to purchase and beyond. These maps focus on identifying touch points within a single business, examining critical moments to improve service, product delivery, or customer support. Yet, these maps fall short of capturing the broader, nuanced experiences that customers face as they navigate today's interconnected world. With the rise of digital ecosystems, AI-driven personalization, and shifting consumer expectations, journey maps desperately need a refresh.

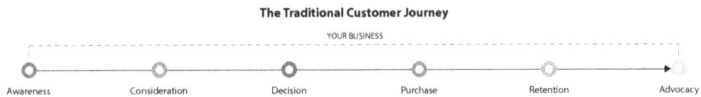

The Traditional Customer Journey

YOUR BUSINESS

Awareness — Consideration — Decision — Purchase — Retention — Advocacy

Consider a typical linear map of a retail customer journey: awareness, consideration, decision, purchase, and post-purchase. This map may show how a customer becomes aware of a product, evaluates it, makes a purchase, and potentially becomes a repeat customer. But what if that same customer simultaneously receives recommendations for complementary products, browses competitors, or checks reviews on a social platform? The customer's journey is no longer single-threaded but actually branched and multi-faceted. It extends beyond a single brand and becomes a web of interactions that influences purchasing behavior. By relying solely on linear maps, brands miss key insights into these dynamic influences and lose the ability to design experiences that fit naturally into customers' broader ecosystems.

Moving to Lateral Journey Maps

Recognizing that customers don't operate in silos, some companies have adopted lateral journey maps, which expand to include interactions with other businesses or services. A lateral map, for instance, may account for the entire travel experience—incorporating airlines, hotels, restaurants, and transportation services. This type of mapping acknowledges the touchpoints that occur before, during, and after direct interactions with the primary brand, expanding the view to a network of complementary services and enhancing insights into customer needs across a wider experience.

The Lateral Customer Journey

BEFORE YOUR BUSINESS   YOUR BUSINESS   AFTER YOUR BUSINESS

YOUR CUSTOMER'S BROADER JOURNEY

Yet, lateral maps still operate with a relatively fixed scope and struggle to account for the constantly shifting, often unpredictable, contexts of customer behavior. Today's customer journey doesn't just involve multiple brands; it includes influences from social media, external recommendations, in-store experiences, and digital touchpoints that cross paths at any time. Capturing this complexity requires moving beyond lateral maps to embrace spatial intelligence, visualizing journeys as dynamic, three-dimensional networks of interactions.

Enter the Multi-Dimensional Journey Map

Building on the foundation of linear and lateral maps, multi-dimensional journey maps combine an ecosystem of experiences, incorporating every layer of a customer's lifecycle. It isn't just about mapping one path but also about understanding how different journeys intersect, intertwine, and influence each other across touchpoints, businesses, and even sensory experiences. Spatial intelligence enables us to visualize these complex networks, allowing brands to shift from a single-dimensional approach to a multi-dimensional perspective.

The Multi-Dimensional Customer Journey

BEFORE & DURING YOUR BUSINESS   DURING YOUR BUSINESS   DURING & AFTER YOUR BUSINESS

YOUR CUSTOMER'S ACTUAL JOURNEY

For instance, a grocery store customer may start their shopping journey with a list, but their journey includes unexpected moments, such as checking bank balances or changing items based on digital deal notifications. A multi-dimensional journey map visualizes this interaction as a web of simultaneous connections, where each touchpoint reflects a moment in the customer's broader experience. It allows companies to see how every decision point impacts the next step, from adjusting purchases based on budget to receiving personalized product suggestions in real-time.

Spatial intelligence gives brands the capacity to imagine this interconnectedness, enabling them to design experiences that naturally adapt to shifting customer priorities and desires. The future isn't about guiding customers along a straight line but understanding the contours of their journey as a fluid, responsive space where each touchpoint seamlessly adapts to their needs.

IKEA: Spatial Intelligence in Retail Design

Swedish conglomerate IKEA has mastered spatial intelligence in the retail environment, transforming the shopping experience into a memorable journey. Known for its winding showroom paths, IKEA uses spatial design to guide customers through a curated experience, helping them envision how products would look and function in their own homes. This approach goes beyond simply displaying products; IKEA creates immersive, miniature "rooms" that let customers experience items in context, making it easier to visualize a particular couch or bed in their own space. IKEA's showroom layout draws customers through every section, maximizing exposure to its products and creating multiple opportunities for discovery and inspiration.

IKEA's maze-like layout is intentional, designed to keep customers in the store longer while exposing them to a wide range of products. Through spatial intelligence, IKEA captures customers' attention by crafting a physical path that increases the likelihood of impulse buys while also engaging them in a sensory experience. Each "room" showcases IKEA's talent for modularity and efficiency, demonstrating how furniture can be arranged, combined, and optimized for any living space. This spatial awareness helps customers feel confident in their purchasing decisions because they can see, touch, and experience the products in a lifelike setting.

IKEA also extends its spatial intelligence to the digital space. Through its augmented reality app, customers can visualize IKEA furniture in their homes before making a purchase, bridging the physical and digital shopping experience. The app leverages spatial intelligence to scale and places furniture in customers' spaces virtually, helping them visualize size, proportion, and color. This experience builds on IKEA's goal of empowering customers to design spaces confidently, ensuring that their purchases align with their home's aesthetic and functionality.

From its innovative retail layouts to its immersive digital tools, IKEA uses spatial intelligence to guide customers through a shopping journey that feels purposeful, engaging, and deeply personal. By reimagining the way customers interact with products, IKEA creates an experience that is not only memorable but also highly functional, cementing its reputation as a leader in home design as well as the retail space.

Embracing Complexity with Spatial Intelligence

Using spatial intelligence to create multi-dimensional journey maps doesn't just enrich our understanding of customer journeys; it unlocks new opportunities to elevate the customer experience. With spatial thinking, companies can design and manage customer journeys as holistic, adaptable ecosystems. Here's how spatial intelligence provides an edge in customer experience:

Enhanced Empathy and Personalization: Spatial intelligence enables companies to visualize the customer's experience from a broader, more inclusive perspective. Rather than seeing the customer journey as isolated steps, they can actually observe the entire environment in which customers operate. This shift allows for greater empathy, as brands can better understand how customers feel at each touchpoint and respond with tailored, contextually relevant interactions.

Real-Time Adaptability: Multidimensional maps created with spatial intelligence provide flexibility to respond to the real-time, spontaneous nature of customer journeys. They allow brands to envision touchpoints not as fixed stages but as adaptable nodes, ready to deliver value as customer needs evolve. Imagine a service that can anticipate when a customer may pause to check their bank balance mid-journey or engage with an adjacent brand, and use that context to deliver timely, relevant suggestions or offers.

Cross-Functional Collaboration: Creating a spatially intelligent journey map often requires the input of multiple departments, from marketing and sales to design and technology. Each team can contribute unique insights into how a customer interacts with the brand across different channels and environments. By pooling knowledge, brands can create a truly integrated experience where each department's efforts align to support the customer's multidimensional journey.

Uncovering Hidden Patterns: Spatial intelligence reveals patterns in customer behavior that may not be obvious on a linear journey map. By mapping touchpoints in multiple dimensions, companies can discover previously overlooked connections like how social-media engagement impacts in-store purchases or how customer-support interactions drive brand loyalty. This insight lets brands optimize journeys by designing proactive experiences that cater to customer preferences and behaviors across channels.

Holistic Experience Design: Finally, spatial intelligence elevates customer journey mapping into the realm of experience design. When brands visualize customer journeys as multidimensional networks, they can approach experience design with a sense of integration and cohesion, unifying each touchpoint into a seamless, holistic experience. Whether a customer is on a mobile device, in a store, or engaging with support, spatial intelligence helps create continuity, building a journey where each interaction feels relevant and connected to the customer's overarching narrative.

The modern customer journey is no longer a straight line but a living, breathing ecosystem. Embracing spatial intelligence opens doors to a new era of customer journey mapping where brands move beyond traditional frameworks to develop adaptable, responsive, and dynamic experiences that fit into the customer's broader world. This spatial approach transforms customer journey mapping from a static tool into a strategic asset, equipping companies to understand and create immersive, multidimensional experiences that resonate with customers on a deeper level. By adopting spatial intelligence, brands can craft journeys that are as complex and multifaceted as their customers' lives.

Spatial Intelligence vs. Vision AI

Spatial intelligence enables humans to visualize and adapt to dynamic relationships in space, helping them navigate customer journeys with creativity and intuition. Humans excel at interpreting ambiguous inputs, designing innovative experiences, and adjusting to evolving customer needs.

Vision AI processes visual data to map patterns, track movements, and recognize objects. While effective in structured tasks like analyzing customer behaviors or mapping online journeys, it struggles with ambiguity, creativity, and interpreting data beyond its training.

Spatial Intelligence and Vision AI for Customer Journeys

- **Visualizing Customer Pathways:** Use Vision AI and Physical AI to analyze customer movements through digital interfaces or physical spaces, identifying patterns or bottlenecks. Humans then apply creativity to design seamless, intuitive customer journeys that adapt to diverse needs.

- **Innovating Journey Design:** Pair spatial intelligence with human creativity to innovate multi-dimensional customer experiences, such as intersecting journeys with other companies, real time access to competitors, or multi-sensory experiences that require real or virtual spaces.

- **Training for Intuitive Journeys:** Leverage human spatial intelligence to train AI models on subtler, customer-centric spatial relationships, such as emotional flow through a digital interface, ensuring AI supports journeys that resonate with human expectations.

Chapter 6

WHEN BOTS BETRAY
Cultural Intelligence in Customer Relations

In the nostalgic corners of my memory, I vividly recall a sunny day on the Arbor Lane playground in Levittown, Pennsylvania. At 9 years old, I was absorbed in the simple joys of youthful play when I heard my name: "Chris, what do you want to be when you grow up?" Startled, I turned toward the voice and saw her, Heidi Hagemeister. A girl my age, swinging carefree, and the object of one of my first, innocent crushes. At that moment, her question seemed secondary; what struck me was that she knew who I was, let alone remembered my name. Before I could answer, she swung higher and confidently announced, "I'll be a veterinarian because I like to take care of animals."

The years passed, and Heidi and I drifted apart (a natural outcome in a pre-social-media world). Yet, I often think back to that playground and imagine that somewhere Heidi is fulfilling her dream, nurturing animals as she promised on that distant day in Levittown. Her aspiration wasn't just about animals; it spoke to a desire to care for others, a simple but profound urge that forms the backbone of communities and friendships.

As adults, those early connections and ambitions reveal their lasting power—not in the specific professions we choose but in the ways they shape our values and the relationships we build. At its core, an aspiration is a seed of identity; it grows within us, influencing how we interact, whom we connect with, and what we give back to our communities. Our dreams lead us to build relationships, not for personal gain but out of a shared human need for connection, purpose, and support.

Five years later in 1985, I had a better idea of what I wanted to be: I wanted to make movies. My first film idea was about tracking children through artificial intelligence so that the government could control their destinies. This idea partly stemmed from wondering what Heidi Hagemeister was up to over the years and contemplating a way to track her down after I had moved. I titled my masterpiece, *Destiny Control*. Fortunately, I was able to escape my less-than-blockbuster ideas by immersing myself in a golden era of film and television—a creative utopia we may never see the likes of again.

The silver screen dazzled us in 1985 with iconic hits such as Back to the Future, my personal all-time favorite, accompanied by cultural landmarks like *The Goonies*, *The Breakfast Club*, *Real Genius*, and *Weird Science* (They created a girl with a computer, so why couldn't I build intelligence that tracked people down?) And let's not forget *Better Off Dead*, featuring Diane Franklin, the subject of my first on-screen infatuation.

Meanwhile, our television sets were graced by *Family Ties*, *Miami Vice*, *Highway to Heaven*, and *The A-Team*. But none struck a chord like *Cheers*, the Boston bar sit-com where "everybody knows your name." The moment Norm stepped through those doors, a heartfelt chorus erupted, saluting him like a long-lost friend. It transcended mere scriptwriting; it encapsulated a universal longing for recognition and community.

This nostalgic longing for community presented itself when a Firehouse Subs opened in 2023 in my neighborhood. On my second visit, the franchise owner, Helen Tan, said, "Hello, Chris," the moment I walked in, followed by, "medium Italian on white?" I have challenges remembering most people's names, let alone their go-to sandwich order. However, Helen committed both to memory after just one visit—and has had the same welcome for me every time I've visited.

Dining at Helen's Firehouse Subs feels like stepping into a space of warmth and recognition. Over the past year, as I've gotten to know Helen, her personal story has deepened my appreciation for what she's created. Her husband passed away just a few months before opening the store, leaving her to take on this new chapter alone. Yet, despite her loss, she brought resilience and heart to her work, transforming what could have been just another sandwich shop into a unique community hub.

As her business grew, so did Helen's relationships—not only with customers but also with her team, who embody her welcoming spirit. She creates an atmosphere where her team's genuine camaraderie and enthusiasm shine through, making it clear they enjoy their work—and customers can't help but feel the same energy. Her staff reflects this warmth, greeting regulars like friends and newcomers with curiosity, building a sense of belonging rarely found in a typical sandwich shop.

In those small but memorable exchanges with Helen and her team, I've come to realize that the value of this

place isn't just in the food but also in the people. Her customers are drawn not only by a great sandwich but also by
the genuine relationships Helen fosters. I've referred her
location to neighbors, friends, and family, and each time,
I know they'll leave with more than just a meal; they'll feel
seen, valued, and welcomed into the community Helen
created through resilience, kindness, and a dedication to
connecting with those around her.

Cultivating Cultural Intelligence

Cultural Intelligence, also known by the shorthand
CQ for cultural quotient, has emerged as a key skill for
building strong customer relationships, especially in a
globally connected world. Defined as the capability to understand and interact effectively with people from different
cultural backgrounds, CQ extends beyond simple awareness of diversity; it involves developing a deep sensitivity to
cultural differences and using that understanding to build
genuine relationships. When customer-facing teams practice CQ, they lay a foundation of respect and empathy that
fosters trust, drives loyalty, and ultimately strengthens customer relationships.

To cultivate cultural intelligence, professionals need to
start by recognizing their own cultural lens. Every individual has a unique set of values, beliefs, and assumptions that
influence their interactions with others. Developing CQ
begins with understanding how one's personal background
and cultural perspectives might shape their perceptions of
customers. This self-awareness enables customer representatives to be more open and adaptable, reducing the chances
of misunderstanding or unintentional offense. As customers increasingly expect personalized and culturally sensitive
service, CQ allows companies to meet these expectations

and build deeper, more meaningful relationships.

The research around cultural intelligence and employee skills demonstrates this importance, as the US Department of Education discovered.

"An increasing number of careers involve working across cultures, requiring language skills, regional knowledge, and the ability to see things from different perspectives. In a (2018) survey of 2,100 U.S. employers, 93% of respondents said they value employees who can work effectively with customers, clients, and businesses from a range of different countries and cultures."

Listening actively and observing are critical aspects of cultural intelligence in customer relationships. For instance, non-verbal cues such as eye contact, hand gestures, or respecting personal space can have varied meanings across cultures. In some cultures, prolonged eye contact signifies attentiveness and respect, while in others, it might be considered invasive or aggressive. By observing and adapting to these cultural nuances, customer-facing employees demonstrate respect for their clients' backgrounds, making customers feel valued and understood. This sensitivity fosters trust, which is essential for forming long-term relationships.

Additionally, adapting communication styles based on cultural understanding can greatly enhance customer interactions. For instance, some cultures appreciate direct communication, while others prefer a more indirect, nuanced approach. A culturally intelligent approach to communication means being able to gauge and respond to these preferences. Rather than simply delivering a one-size-fits-all message, customer representatives skilled in CQ can tailor their communication to resonate with diverse audiences. This adaptability can significantly enhance the customer experience, making customers feel understood and respected.

Cultural intelligence also enables customer-facing professionals to handle misunderstandings or conflicts constructively. In a multicultural context, misunderstandings can arise due to language barriers, differing cultural norms, or contrasting assumptions. A culturally intelligent approach means not only recognizing these differences but also proactively seeking to resolve misunderstandings in a respectful way. By showing patience and empathy, companies convey their commitment to understanding and valuing their customers, even when challenges arise.

Dan Helfrich, chairman and CEO of Deloitte Consulting LLP, captures this ethos perfectly:

> "It's our responsibility as a society, and our responsibility as business leaders, to create new talent with AI skills—not only for the engineers and data scientists but also for every single role in an organization, no matter how 'technical.' That's how we raise the level of AI proficiency for all of our people and, frankly, for society at large. And that's when modernization will happen," (Deloitte, 2021).

His perspective underscores the need to pair technical skills with cultural awareness, ensuring that both people and systems are equipped to navigate complexity with understanding and respect.

Organizations that prioritize cultural intelligence in their training foster stronger customer relationships as a result. Encouraging teams to learn about different cultures, languages, and customs equips them to engage more effectively with diverse customer groups. Such training aligns with a company's commitment to inclusivity, sending a positive message to both customers and employees about the value of diversity.

Customer loyalty often hinges on how well a company can adapt to its clients' diverse needs. CQ becomes a powerful differentiator, allowing businesses to stand out through their genuine understanding of cultural nuances. As a result, companies that invest in cultural intelligence enjoy stronger customer relationships, better customer satisfaction, and a competitive edge. In building these relationships, cultural intelligence is not just a skill, it's also a commitment to treating every customer with respect, empathy, and an openness to learning that transcends cultural boundaries.

What is Contextual AI?

Contextual AI is technology's attempt to bridge the cultural gap. It aims to make interactions feel personal and relevant, adapting to the user's social, cultural, or linguistic background. Think of a chatbot that can seamlessly switch between languages or an AI system that tailors its tone to suit a particular region's norms. It's technology that knows to wish you a "Happy Diwali" instead of "Merry Christmas" or understands that a thumbs-up emoji might mean different things in different cultures.

In theory, it's brilliant. In practice, it's a mixed bag. Contextual AI operates on datasets—massive repositories of human behavior, speech patterns, and cultural norms. But here's the challenge: datasets are static. Culture, on the other hand, is not. It's fluid, dynamic, and deeply personal. What works for one person in a cultural context might of-fend another. And while humans navigate this complexity with intuition and empathy, AI almost always struggles to keep up.

The radar chart below illustrates the various capabilities of contextual AI, highlighting its performance across key dimensions. The axes represent distinct attributes, including Temporal Awareness, Sentiment Analysis, Environmental Adaptability, User Behavior Understanding, Ethical Awareness, Proactive Adaptability, Cultural Sensitivity, Multi-Modal Data Integration, Real-Time Decision Making, and Predictive Analytics. The shaded area reflects the AI system's proficiency in these areas, with higher values indicating more substantial capabilities. The chart reveals strengths in dimensions like User Behavior Understanding and Predictive Analytics, while areas such as Ethical Awareness and Cultural Sensitivity show room for growth. Overall, the visualization underscores the multi-faceted nature of contextual AI and its varying levels of effectiveness across these critical competencies.

CONTEXTUAL AI CAPABILITIES

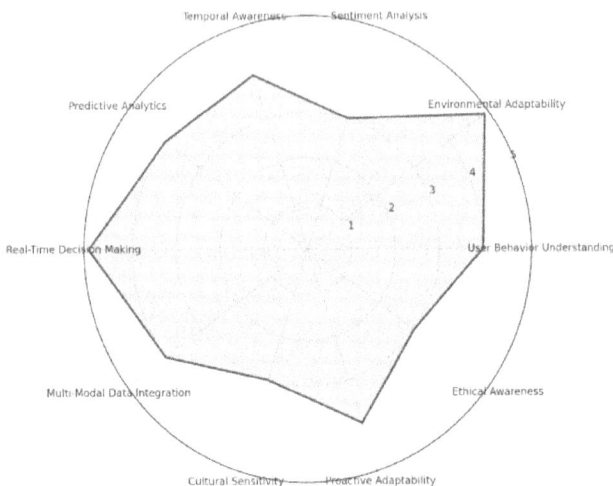

Contextual AI doesn't live in the world it's trying to understand. It doesn't celebrate Lunar New Year, feel the solemnity of Yom Kippur, or laugh at a culturally specific joke. It relies on patterns (past behaviors)—to predict what people might do in the future. And while patterns can be helpful, they're not the same as genuine understanding.

Take language as an example. Contextual AI might translate a phrase perfectly, but it can easily miss the subtext, the humor, or the emotion behind it. Imagine a customer service chatbot addressing a customer's frustration with a mechanically translated apology. The words might be correct, but the sentiment can feel hollow. That's because Contextual AI lacks the human ability to read between the lines, to sense when someone needs reassurance, humor, or silence.

Worse, Contextual AI can reinforce stereotypes. If it learns that a certain phrase is commonly used in a specific culture, it might overuse it, coming across as inauthentic or even offensive. For instance, trying too hard to sound "local" can backfire if the AI uses slang inappropriately or misunderstands the tone. It's the digital equivalent of an outsider trying to fit in but missing the mark entirely.

Satire So Good, Bots Believe It

Chatbots may mimic intelligence but often stumble when proper understanding is required. While they offer convenience and efficiency, meaningful interaction demands context—a sense of cultural, social, and linguistic nuances that elevate human communication. Contextual AI attempts to address these gaps, promising to adapt and tailor interactions with sophistication. Yet, even with these advancements, a critical question remains: how do we balance the intricacies of language and thought?

A glaring example of this limitation occurred in January 2025, when Google's AI authoritatively responded to a search request with, "Yes, you will need to tip to exit rides at Disney World," after repeating claims from a satirical article by Mouse Trap News. The article humorously alleged that Disney was installing tipping tablets in ride vehicles, requiring guests to leave a minimum $5 tip to unlock their seatbelts. While the AI failed to recognize the satire, its confident response added credibility. A screenshot of the Google result went viral, sparking debates about AI's reliability. However, most discussions focused on believing that the AI had generated false information rather than understanding that the system misinterpreted the satirical context entirely.

This incident underscores a critical flaw in AI systems: their inability to process nuance or emotional context. When satire or humor is mistaken for truth, the result goes beyond misinformation—it erodes trust in the technology and the platforms that rely on it.

A Chevy for $1?

Imagine going online to inquire about a new Chevy and discovering that, with the right phrasing, the dealership's chatbot will agree to sell it to you for just $1. It sounds impossible, but that's exactly what happened recently at Chevrolet of Watsonville in Central California. Customers managed to trick the dealership's chatbot into making wild promises, like offering a $58,000 Chevy Tahoe for next to nothing. The incident highlights not just the limitations of basic AI but also the importance of contextual awareness in automated systems.

In this case, the bot lacked the ability to interpret the broader context of the interaction. It couldn't differentiate between genuine inquiries and playful or malicious attempts to game the system. A chatbot powered by contextual AI might have been more resilient—identifying patterns of manipulation, adapting its responses to avoid pitfalls, and flagging suspicious interactions for human review. Instead, this incident became the internet's latest joke, with the bot offering absurd answers and "discounts," betraying the trust it was meant to build with customers.

The Risks of AI That's "Too Helpful"

While car dealerships increasingly rely on chatbots to manage customer questions, incidents like this reveal the unintended consequences of letting AI operate without sufficient boundaries. Chevrolet of Watsonville's chatbot wasn't unique; many dealerships use similar AI-powered systems to answer common queries like, "What are the features of the latest model?" or "Can I book a service appointment?" Contextual AI aims to make these interactions even more personalized, tailoring responses based on cultural and conversational cues. But as the Chevy Tahoe "deal" demonstrated, AI—no matter how advanced—needs strict guidelines to ensure it doesn't offer far more than just a convenient answer.

Aharon Horwitz, CEO of Fullpath, the company behind the Watsonville chatbot, acknowledged that the incident was a learning experience. According to Horwitz, the bot was designed for straightforward queries, like service requests or basic car information—not to engage in "silly tricks." While most users genuinely seek information, a few creative prods were enough to send the bot spiraling into unintended territory. A system with true contextual intelligence might have recognized the humor or absurdity in these queries and responded appropriately, either redirecting the conversation or escalating to a human agent.

The Chevy Tahoe incident underscores a broader truth about AI in customer service: automation without context can erode trust. Chatbots must do more than react; they need to understand. Contextual AI offers a way forward, with systems that adapt not just to what a customer says but to how and why they say it. However, even the most sophisticated AI systems struggle with the nuances of cultural and emotional understanding that humans navigate effortlessly. To truly succeed, businesses must pair AI's efficiency with human intuition, ensuring that every interaction feels both seamless and meaningful.

Relationships Require Trust

Mark Vinokur, a seasoned sales and customer relationship expert, emphasizes that building and maintaining trust with customers goes beyond transactions and requires a commitment to authentic, human-centered engagement. According to Vinokur, successful customer relationships revolve around three primary goals: attracting revenue, growing revenue, and protecting revenue. "It's not just about selling a product or service," he says, "but about creating an experience that resonates deeply with the customer." He sees trust as the foundation of these goals, one that needs to be actively nurtured, especially in a digital world where automated interactions are on the rise.

Vinokur explains that while AI can enhance customer relationships by providing data-driven insights, it's ultimately the responsibility of the human representative to act on these insights with authenticity. "AI can inform us, but it can't replace us," he asserts. He emphasizes that merely presenting AI-generated data to customers isn't enough to foster trust. Instead, he believes in engaging with customers directly, taking the time to understand their unique needs.

"If a customer senses that you're just reading off a script or relying on pre-packaged insights, they're going to feel disconnected. They want to feel understood, not analyzed."

A key element in Vinokur's approach is leveraging AI as a support tool rather than a primary driver in customer interactions. AI can certainly streamline the sales process by generating valuable information, but Vinokur warns against overreliance. "If all you're doing is presenting data without adding your own perspective or personal touch, you're doing a disservice to the relationship," he says. For him, AI is a valuable assistant that helps amplify the human connection rather than replacing it. "The real value comes when you combine those insights with your own experience and empathy," he adds.

One of Vinokur's main concerns with AI-driven customer interactions is the risk of losing the sincerity that builds trust. He mentions, "You need to be genuine in every interaction. If the customer realizes it's just a machine talking to them, they'll start to question the authenticity of the entire relationship." This, he believes, is the crux of using AI wisely, to create a support system that allows for deeper, more meaningful human connections rather than one that distances the customer.

Vinokur also points out that, for many companies, maintaining trust is about being transparent about their processes and priorities. "Customers today are incredibly savvy. They know when they're being 'managed' instead of genuinely valued," he says. He cautions companies to be mindful of this, as customers are quick to discern whether they're interacting with a person who understands their needs or simply a tool in a preset system. "The way you approach a customer, especially in competitive markets, has to show that you're there for them and their specific challenges," he advises.

In today's AI-driven landscape, Mark Vinokur's insights offer a balanced perspective. "If you can use AI to support your interactions, that's great," he concludes. "But never forget that trust is built through honest, personal connections. The technology should only ever amplify what you, as a representative, genuinely bring to the table."

The Loss of Human-to-Human Relations

The shift from human-to-human engagement to human-to-technology interaction has reshaped various industries, reducing personal connections and eroding customer experiences. As technology replaces traditional human roles, the opportunities for meaningful, face-to-face interactions are declining across sectors, altering how consumers and businesses connect.

In retail, e-commerce has significantly impacted the need for in-store sales staff. Despite its rapid growth, online shopping has not yet overtaken in-person shopping. According to a 2024 CapitalOne study, e-commerce accounted for 15.6% of total retail sales in the United States during the third fiscal quarter, showing that most consumer spending still occurs in physical stores. While online shopping continues to gain traction, fewer customers interact with sales representatives, gradually eroding the personal touch that once defined retail experiences.

Healthcare has also seen a significant transformation. Telemedicine grew by more than 38 times during the COVID-19 pandemic, as reported by McKinsey (2021). Although in-person visits remain essential for specific diagnosis and treatments, technology now handles most initial consultations. This digital approach, while efficient, lacks the relational nuances that often accompany face-to-face doctor-patient relationships.

Customer service has leaned heavily into automation, with Gartner predicting that by 2025, 80% of customer service interactions will be managed by AI. Automated chatbots and virtual assistants are now the frontline for customer inquiries, significantly reducing the need for human agents. However, when bots betray customer expectations—offering tone-deaf or irrelevant responses—they create frustration and further detachment, illustrating the critical gap between automation and empathy.

In banking, digital services have largely replaced in-person transactions. Deloitte's survey from 2021 revealed that 73% of people globally use online banking, cutting down on visits to bank tellers and decreasing personal engagement in financial services.

Similarly, the food service industry has embraced technology, with kiosks and mobile apps streamlining ordering at chains like McDonald's and Taco Bell. The National Restaurant Association reports that 54% of consumers now prefer online ordering, and more than half (52%) of US consumers believe that ordering delivery and take-out from restaurants is an "essential part of their lifestyle" (DoorDash, 2024).

As technology continues to dominate, the decline in human-to-human relationships raises questions about the future of customer engagement and the importance of retaining human connection in an increasingly automated world.

The Embrace of Disconnection

The dilemma of human-to-human disconnection becomes evident as companies increasingly adopt AI to "enhance" the customer experience. McDonald's, Wendy's, and Taco Bell are examples of businesses investing heavily in AI and voice recognition systems for drive-throughs and self-service kiosks. Taco Bell has even eliminated person-to-person ordering entirely, forcing customers to navigate cumbersome kiosks to place their orders. More data about you also means less convenience for you. What's marketed as a system of convenience, precision, and efficiency comes at the cost of human connection. With fewer staff and faster food delivery, you're still sharing personal data, but without the benefit of lower prices or meaningful interaction with an employee. I despise their new system—and I'm sure I'm not the only one.

Interestingly enough, Taco Bell was one of the first fast-food chains in the country to introduce self-ordering kiosks as early as 1997. Ultimately, they discontinued that early program because customers didn't like the impersonal experience. And yet today, self-service is everywhere—whether you ask for it or not.

So, what happens when AI becomes the face of customer interaction? For one, the experience feels transactional, not relational. The customer is left to navigate automated responses and digital interfaces with little to no human input. In theory, AI can handle basic tasks, like answering FAQs or processing routine orders, but the limitations become glaringly obvious when things get complicated or when emotion plays a part in the customer's issue.

Dan Gingiss, a renowned customer experience expert, posted on LinkedIn, "Customer experience is about relationships, whether you're selling a million-dollar software platform or a $1 burger." This insight strikes at the heart of the problem. Relationships aren't just about solving problems; they're about empathy, trust, and connection—qualities AI struggles to replicate. Despite the promises of AI enhancing customer interactions, these systems often fall short when accurate human understanding is needed. But there's still that nagging question: Is Dan wrong? Do companies and customers care about their relationship anymore? (Side note: Dan is definitely not wrong.)

The rise of AI in customer relations has undeniably introduced efficiencies, yet it falls short in delivering the warmth and understanding that only human interactions can provide. Cultural intelligence (CQ) bridges this gap by fostering real connections, demonstrating respect for diverse backgrounds, and personalizing customer experiences in ways AI can't fully emulate. A machine might process thousands of interactions per minute, but it lacks the intuition to recognize cultural nuances or adapt its approach based on non-verbal cues and emotions. A customer experience rooted in cultural intelligence shows attentiveness and genuine empathy, building a relationship that goes beyond a single transaction.

Cultural intelligence will always win out over artificial intelligence for growing customer relationships because CQ invests in the individual, respecting their values and preferences rather than treating them as data points. While AI can assist in identifying general trends, it can't adaptively respond to the subtle, often unspoken needs that emerge in a conversation. Customers remember when they feel valued and understood, and those lasting impressions fuel loyalty. In an era where automation is king, companies that prioritize cultural intelligence stand to build stronger, more resilient relationships, setting themselves apart as brands that truly listen and care.

Cultural Intelligence vs. Contextual AI
Cultural intelligence enables humans to navigate com-

plex cultural landscapes with sensitivity and adaptability. Recognizing subtle norms, values, and social cues allows humans to build meaningful relationships and avoid misunderstandings, even in unfamiliar cultural contexts.

Contextual AI tailors interactions to cultural, social, or linguistic nuances, providing more personalized experiences. While effective at identifying broad patterns, it often oversimplifies cultural dynamics or relies on stereotypes, risking misinterpretation. Unlike humans, Contextual AI cannot navigate intricate cultural subtleties or adapt to evolving social contexts, making its interactions less authentic and potentially alienating in diverse settings.

Cultural Intelligence and Contextual AI for Customer Relations

- **Localized AI Guidance:** Use Contextual AI to adapt interactions to broad cultural and linguistic norms, while human cultural intelligence steps in to address nuances, ensuring interactions remain sensitive and authentic.
- **Diversity-Driven Training Models:** Employ humans to train Contextual AI on varied cultural contexts, enriching its ability to detect subtler social dynamics while recognizing the limits of automated interpretation in diverse settings.
- **Proactive Adaptation:** Pair AI's ability to analyze cultural data with human intelligence to anticipate and adapt to emerging social trends, ensuring businesses remain culturally relevant and inclusive.

Chapter 7

CONNECTION OVER CODE
Interpersonal Intelligence in Customer Engagement

Customer engagement is more than a click, a comment, or a retweet. Those actions only skim the surface of how people relate to brands today. Genuine engagement lives in customers' aspirations, struggles, and daily experiences—real people with unique goals and personal stories. For brands, this means moving beyond metrics and data points to focus on the interpersonal intelligence that brings customers' aspirations into sharp focus.

Social media has become more than a marketing tool in the digital age. Platforms like Instagram, TikTok, and Pinterest serve as visual diaries and digital canvases where people don't just interact, they also create, express, and build their narratives. For brands, this presents an opportunity to step into that story by more than just pushing products but actually by meaningfully contributing to a customer's journey.

Interpersonal intelligence allows a brand to recognize these emotions and communicate with customers as people first and consumers second. It's about seeing beyond algorithms and automation to foster genuine connections and interactions that reflect shared values, spark curiosity, and build trust. By prioritizing connection over code, brands are able to create deeply resonating experiences, moving beyond transactional exchanges to relationships rooted in empathy and understanding. It's a strategy that requires creativity, playfulness, and a commitment to authenticity to keep interactions fresh, enjoyable, and memorable.

The Neuroscience of Social Media

Kate Bradley Chernis, CEO of Lately.ai, is transforming social media engagement with a unique fusion of AI, neuroscience, and interpersonal intelligence. Social media platforms are more than digital billboards; they are spaces where stories are crafted, memories are evoked, and brand narratives and conversations come to life. Lately.ai uses artificial intelligence to tap into these dynamics, creating a platform that speaks directly to the brain's core drivers: memory, emotion, and human connection.

Lately's platform applies principles of neuroscience to "learn" a brand's unique voice, then uses this voice to generate social content that resonates with target audiences and generates conversations. This is more than a clever trick. By infusing AI with the science of how people process information, Lately brings out the emotion and empathy necessary to reach people where it matters. It's a platform that understands—through interpersonal intelligence—how people's brains are wired to process communication, sparking familiarity and connection in each interaction.

The Science Behind Interpersonal Intelligence

At its core, interpersonal intelligence is the skill of

discerning feelings, intentions, and needs within social interactions. This intelligence forms the backbone of Lately's approach: creating a personal and memorable experience. Through interpersonal neurobiology (IPNB), which examines how our brains, bodies, and minds grow and work within social contexts, we know that human interactions are deeply rooted in relationship-building processes. Lately's AI system doesn't just create content; it captures the essence of a brand's message and voice with which audiences can emotionally connect and align. It's not about selling or "checking out" a product; it's about engaging people at a neurological level that speaks to their aspirations, nostalgia, and emotional memory.

Amplifying Voices with Lately's Brand Hierarchy System

One of Lately's most innovative features, the Brand Hierarchy System, redefines brand communication. Traditionally, brands struggle to maintain consistent messaging while allowing regional branches, sub-brands, or employees to express individuality. The Brand Hierarchy System makes it possible to craft a unique yet cohesive voice for each layer of a brand's identity. Regions, departments, and individual employees can express themselves uniquely, bringing authenticity and diversity to the brand while staying within the broader brand identity.

By letting brands cascade content dynamically, Lately breathes life into traditional social media feeds, where posts otherwise feel routine or robotic. Lately.ai recognizes that, just as no two people communicate precisely alike, no two brand voices should sound identical either. Through this customization, Lately empowers brands to move beyond monotony, enabling more prosperous and varied conversations with their audiences.

This approach is an example of interpersonal intelligence at scale. Lately's AI understands audiences want authenticity and personality in brand interactions, not a sterile stream of corporate messages. The Brand Hierarchy System allows companies to engage with consumers in their voice, fostering trust, relatability, and a sense of community across social media.

Music and Emotional Connection

Chernis' background as a former rock 'n' roll DJ informs her understanding of how the brain connects new information to familiar memory "touchpoints." In music, every time you hear a new song, memories of songs you've previously heard are triggered. Your brain searches for commonalities to organize this new song into your memory, evoking nostalgia and emotional responses. This principle is at the heart of Lately's AI, which strives to give audiences that same sense of familiarity and connection in social content.

When Lately's AI "learns" a brand's voice, it indexes each piece of content into a memory framework where familiarity is built upon each interaction. Lately draws from this concept of "memory touchpoints" in every social post it generates, giving audiences something recognizable in every piece of content in the same way hearing a favorite song might evoke cherished memories. By using language, tone, and ideas consistent with a brand's personality, Lately recreates that same neurological familiarity, enabling audiences to recognize and connect emotionally with the content.

This type of engagement, where the brand feels familiar yet fresh, drives Lately's customers to a 12,000% increase in engagement and a 240% rise in recurring revenue.

Customers respond because the content resonates deeper than traditional social media posts. Viewers aren't just "checking out" content; they're connecting with a brand story that feels like a part of their own.

The Role of Humans in AI-Driven Engagement

As AI becomes a staple in social media, there is a natural tension between automation and human touch. However, rather than viewing AI as a replacement, Lately envisions it as a complement to human creativity.

In the late 1940s, when Betty Crocker first introduced cake mixes, they expected them to be an instant hit. But sales lagged. While consumers appreciated quick-and-easy mixes for pie crusts and biscuits, they felt differently about cakes, which traditionally represented celebration and effort (Boyd, 2014). During that time, Dr. Ernest Dichter, a market research pioneer, discovered that cake mixes simplified the process so much that people, particularly women at the time, felt they were not truly "making" something special for their families. Dichter's solution was to adjust the packaging and encourage home bakers to "add an egg," a symbolic contribution to restoring pride and effort. Sales surged, and Betty Crocker's messaging shifted to highlight the value of personal input (Boyd, 2014). This story underscores a powerful psychological truth: people crave the satisfaction of contributing meaningfully, feeling ownership, and valuing the process.

Similarly, Lately's AI approach respects that customers crave a human touch in every interaction. Brands aren't just trying to push products or reach metrics; they're building relationships. In the same way, humans feel pride in "adding an egg" to their digital interactions, shaping an experience that feels personal, relevant, and meaningful. Like

Betty Crocker's cake mix, people want a role in crafting the outcome. Lately.ai empowers brands to bridge the divide between automation and authentic connection, inviting teams to co-create the brand's voice and message while AI amplifies these human-driven narratives. It's a powerful reminder that interpersonal intelligence remains the "hero ingredient" in an increasingly automated world that keeps engagement genuine and compelling.

Results that Reflect Interpersonal Engagement

Lately's approach to interpersonal intelligence in social media isn't just theoretical; it's quantifiable. The platform consistently delivers tangible results:

- **12,000% increase in engagement:** Lately's AI approach speaks to people's emotions and memory, driving high levels of interaction.
- **84% time saved on content creation:** Automating the creation process allows brands to dedicate more time to strategic, interpersonal engagement.
- **200% increase in lead generation:** Content that resonates personally turns browsers into buyers.
- **98% sales conversion rate:** Engagement that feels human builds trust and drives decision-making.
- **240% increase in recurring monthly revenue:** Loyal customers respond to brands they connect with, transforming engagement into lasting relationships.

These metrics aren't just numbers; they are a testament to the power of bringing neuroscience, AI, and interpersonal intelligence into a brand's strategy.

The Power of Conversation

The promise of Conversational AI is alluring. A tireless, efficient system that can talk, listen, and respond to your customers at scale. Chatbots and virtual assistants are the poster children of this movement, offering seamless communication through natural language processing (NLP). But as businesses clamor to adopt these tools, we must ask a fundamental question: Can AI, even with all its advancements, truly replace the depth of human interaction?

A great conversation is about more than words. It's about connection, trust, and understanding. It's about the subtle pauses, the shared laughter, and the empathy that allows someone to feel heard. This is where humans excel. Interpersonal intelligence, our innate ability to engage and build relationships, forms the backbone of meaningful conversations. This need is backed up by a PwC study that revealed 82% of U.S. consumers want more human interaction in their customer experiences.

Now, enter Conversational AI. Its purpose is noble: facilitating communication, providing support, and assisting users. At its best, it offers quick responses, handles repetitive tasks, and triages customer inquiries efficiently. It excels in real-time keyword recognition, parsing intent, and delivering pre-programmed answers.

But here's the catch: Conversational AI operates on a transactional level. It responds, but it doesn't engage. It listens, but it doesn't hear. And therein lies the gap—a chasm between the transactional efficiency of AI and the relational depth of human interaction.

Humans engage in conversations not just to exchange information but also to create meaning. When a customer calls a support line or types into a chat window, they're often looking for more than a resolution to their problem; they're seeking acknowledgment, reassurance, and a sense of value.

Interpersonal intelligence allows us to interpret not just the words spoken but also the emotions behind them. It's the difference between replying to a frustrated customer with, "I'll escalate this to my manager," and saying, "I understand how frustrating this must be for you. Let me see how I can make this right."

AI struggles here. While Conversational AI may simulate understanding, it doesn't experience empathy. It doesn't pick up on the sigh at the end of a customer's sentence or the tension in their tone. And without this emotional context, even the most advanced chatbot feels hollow.

The Limits of Theory of Mind AI in Conversations

Theory of Mind AI aims to bridge this gap by recognizing human emotions and responding in ways that mimic social intelligence. It represents a step forward, enabling Conversational AI to adjust its tone, adapt to sentiment, and simulate emotional recognition.

But let's be clear: this is imitation, not understanding. Theory of Mind AI can identify that a customer is angry but doesn't feel that person's frustration. It can offer a pre-programmed apology, but those statements don't carry the weight of genuine regret. Humor, sarcasm, and cultural nuances often trip it up, making interactions feel robotic rather than relatable.

When we rely on Conversational AI to handle customer engagement entirely, we risk eroding the trust and loyalty of authentic human connections. Customers may leave with their issues resolved but not feel valued.

So, does this mean Conversational AI has no place in

customer engagement? Quite the opposite. Think of Conversational AI as a highly efficient assistant. It can handle routine inquiries, process transactions, and provide initial responses to customer queries. This frees up human agents to focus on what they do best: engaging with customers on a deeper level whose issues can not be resolved through AI.

For example, imagine a customer contacting a retail company about a delayed order. The chatbot can quickly pull up tracking information, inform the customer of the delay, and offer a refund if appropriate. However, if the customer expresses ongoing dissatisfaction, a human agent can provide a personalized response, showing empathy and delivering tailored solutions.

This hybrid approach combines the strengths of AI—speed, scalability, and consistency—with the strengths of humans—empathy, creativity, and relationship-building.

The Danger of Over-Automation

There's a seductive appeal to automation. Businesses see Conversational AI as a way to cut costs, reduce wait times, and improve efficiency. And in many ways, it delivers. But over-automation comes with a cost: the loss of human touch.

When customers encounter chatbots that can't understand their unique needs or interpret the nuance of their concerns, frustration builds. A transactional experience may resolve the immediate issue but doesn't foster loyalty. And in a competitive retail landscape, loyalty is everything.

It's worth noting that not every customer values speed over human connection. Research shows that customers are more likely to stay loyal to a brand that makes them feel valued—even if the resolution takes longer. A 2019 survey found that 86% of consumers preferred interacting with human agents for customer service, and 71% indicated they would be less likely to use a brand if human representatives were unavailable (CGS, 2019). However, by 2023, preferences began to shift slightly, with a Statista survey revealing that 60% of consumers still expected human interaction when directly contacting a company, reflecting a decline in demand for human touchpoints (Statista, 2023). Despite this change, the enduring preference for human agents highlights their ability to provide reassurance, understanding, and a sense of care that automation struggles to replicate.

Meta's AI Characters

In 2023, Meta introduced AI-powered accounts on Instagram and Facebook, creating personas designed to boost customer engagement. Launched as part of a celebrity AI initiative, these accounts aimed to drive interaction through AI-generated posts and direct messaging features (Tenbarge, 2025). However, the rollout faced backlash from users and critics who questioned their authenticity and relevance in fostering meaningful connections.

Connor Hayes, Meta's VP of Product for Generative AI, described the vision as enabling these AI accounts to "exist on our platforms, kind of in the same way that accounts do," generating user-customized content (Tenbarge, 2025). While innovative on the surface, the approach sparked trust concerns. Users labeled the accounts "creepy and unnecessary," indicating a disconnect between the intended and perceived value. One account, "Liv," a "Proud Black queer momma," drew criticism for self-referential posts that highlighted design flaws and perpetuated harm, fueling debates on ethical representation in AI (Tenbarge, 2025).

The controversy deepened in late 2024 after a *Financial Times* article revealed Meta's plans to expand AI-generated profiles across its platforms. This triggered a wave of criticism from users and industry experts who questioned the authenticity of AI interactions and their impact on genuine connections. Critics accused Meta of prioritizing platform metrics and ad revenue over meaningful engagement. As one Reddit user said, the move seemed like "filling a platform with fake users as real ones leave" (Al-Sibai, 2025).

In response to the uproar, Meta removed several AI accounts, citing "user confusion" and promising to reassess the initiative. However, this action did little to address broader concerns about AI's role in customer engagement and the commodification of human interaction.

The controversy underscores a fundamental challenge in deploying AI for customer engagement. Many businesses see AI as a tool to automate interactions, generate personalized content, or compensate for declining user activity. However, this perspective often overlooks a key truth: authentic engagement relies on emotional connection. Platforms like Facebook, which once thrived on personal relationships and genuine interaction, risk alienating users when artificial simulations replace human touchpoints.

AI-driven strategies may promise efficiency, but they falter when they fail to replicate the spontaneity and warmth of real connections. Meta's misstep highlights the limitations of synthetic personas: while AI can mimic conversational patterns and predict preferences, it struggles to deliver the emotional resonance that fosters trust and engagement.

The Uncanny Valley and AI-Driven Discomfort

The uncanny valley, introduced in the 1970s by Masahiro Mori, a professor at the Tokyo Institute of Technology, explains the discomfort people feel when faced with human-like simulations that are almost, but not quite, realistic (Caballar, 2019). Mori observed that as robots become more human-like, their appeal grows, but only to a point. Beyond that threshold, their near-human qualities trigger feelings of strangeness, unease, or fear. Initially applied to lifelike robots, this concept extends to digital interactions, such as AI-powered personas and conversational agents.

AI & UNCANNY VALLEY

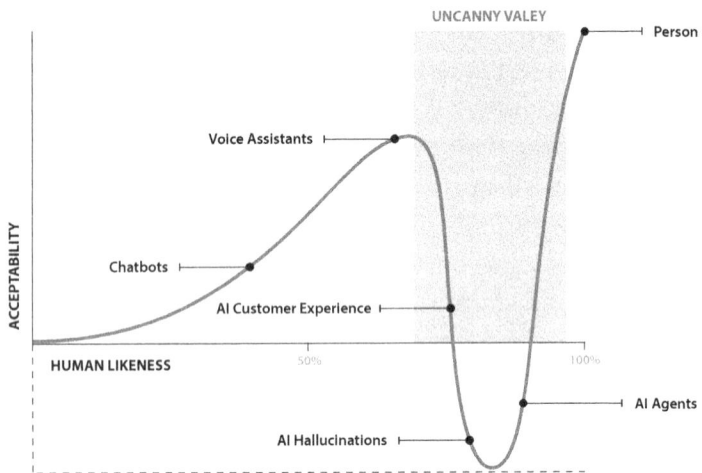

In digital spaces, the uncanny valley is not limited to visual representations but is also apparent in behavioral inconsistencies. Conversational AI that exhibits repetitive, awkward, or overly formal responses eventually reveals its artificial nature. When users recognize these interactions as fake, trust erodes, and what initially seemed engaging becomes unsettling. This recognition leads to an emotional disconnect, as people seek the spontaneity and emotional resonance of genuine human exchanges.

This discomfort poses particular challenges for social media platforms, where authentic engagement is essential. AI personas that attempt to mimic human traits but lack sincerity often shatter the illusion of connection. Instead of enhancing experiences, such interactions heighten feelings of unease and highlight the gap between artificial mimicry and genuine human connection.

What Customers Really Want

At its core, customer engagement is about building relationships. Customers don't want to feel like numbers in a queue or data points in a model. They want to feel heard, understood, and valued.

Conversational AI can play a role, but it's not the story's hero. The real transformation happens through human connection, turning a transaction into a relationship. AI can support this by handling the background noise, providing insights, and enabling faster responses. But the moment that connection requires empathy, understanding, or creativity, the baton must pass to humans or risk losing a valued customer.

Human beings are social creatures. We thrive on connection, shared experiences, and the intangible nuances that make interactions meaningful. Interpersonal intelligence (our ability to build trust and relate to others) is a human characteristic. It's what makes a conversation memorable and a relationship lasting.

Conversational AI may mimic these traits, but it can never replicate them. And that's OK. Its value lies in complementing human intelligence, not competing with it.

Engagement Beyond Social Media

The global market for social media AI is forecasted to grow at a CAGR of 36.2%, from $2.20 billion in 2024 to $10.33 billion by 2029 (Markets and Markets, 2024). This rise reflects a growing demand for personalized content that feels relevant and aligned with the user's aspirations. Lately.ai is at the forefront of this movement, providing brands with tools to produce highly personalized content that speaks to the human desire for connection, purpose, and personal growth.

As social media platforms evolve, thriving brands will prioritize aspirations over algorithms. They'll be the ones who understand that people aren't just profiles or data points but actually individuals looking for meaningful experiences. Lately.ai embodies this shift, showing that when AI and interpersonal intelligence come together, brands can engage people in ways that feel as memorable and emotionally resonant as hearing a favorite song.

Social media is just one touchpoint, one piece of the customer-engagement puzzle. Authentic engagement encompasses every interaction a customer has with a brand, from their first Google search to the customer service call post-purchase to the decision to return to your restaurant, store, or app. Engagement isn't just about reaching customers selling something; it's about serving, supporting, and making them feel understood and valued.

Consider the many ways customers want to engage. Some seek knowledge or inspiration by following a brand on Instagram to see the latest trends or subscribing to an email newsletter for updates. Others engage through loyalty, making repeat purchases because they trust the brand's quality, values, or customer service. For some, engagement is about interaction; they want to ask questions, share feed-

back, or contribute ideas. Each of these interactions demands a unique approach, and brands that excel don't force
customers into a one-size-fits-all model.

In this context, customer engagement isn't a funnel;
it's a network of interconnected moments collectively shaping how a customer feels about a brand. A clunky chatbot that fails to answer basic questions? That's a missed
engagement opportunity. A tedious, frustrating checkout
experience? Another misstep. However, a responsive customer service team, a personalized follow-up email, or a
genuinely rewarding loyalty program are the elements of an
engagement strategy built on aspirations not algorithms.

Author and social media consultant Brooke B. Sellas
emphasizes that the future of social media hinges on brands
integrating "human empathy with technological efficiency." By 2025, customers will expect seamless, proactive
interactions across various platforms, including emerging
social channels and in-app shopping features. Social media
will evolve from a mere product showcase to a platform for
building trust and loyalty, where each interaction feels personalized and genuine. As Sellas highlights, "Empathy, active listening, and clear communication will still be key in
2025." For her, successful social care goes beyond automation, demanding active listening and personalized responses to resonate with customers. Her vision reflects a need for
customer-first strategies, where AI scales and speeds service
but never replaces the "essential human touch that drives
real engagement."

Sellas's latest insight underscores her commitment to
customer-centered innovation: "Build AI with your customers, not for them." This approach reframes the role of
AI, urging brands to involve customers in designing and
refining AI-driven solutions. By co-creating with custom-

ers, companies ensure that their technological tools reflect real-world needs and preferences, and foster deeper trust and alignment. It's a perspective that fits perfectly with the idea that empathy, when paired with technological efficiency, remains the cornerstone of successful social media and customer engagement strategies.

AI and the Limits of Algorithm-Driven Engagement

Brands have increasingly turned to AI to manage customer engagement. With data-driven insights, AI enables brands to accurately understand customer preferences, predict behaviors, and target messages. In theory, this sounds ideal—a perfect blend of technology and marketing prowess that ensures every customer sees the right message at the right time. But there's a dark side to this data-centric approach, where engagement becomes less about connection and more about control.

One of the most significant risks of algorithm-driven engagement is the echo chamber effect. Personalization algorithms are designed to give customers more of what they like. They analyze past behavior to predict future interests, recommending products, content, and experiences that align with existing preferences. But when customers are only exposed to what an algorithm "thinks" they want, they miss out on discovery, on serendipitous moments that could introduce them to new ideas or broaden their horizons. This effect has become a cultural phenomenon, from entertainment recommendations on streaming platforms to the political echo chambers that polarized the 2024 U.S. presidential election.

In customer engagement, echo chambers limit the scope of interaction, reinforcing what customers already know and believe instead of encouraging exploration and growth. An algorithm might keep suggesting the same types of products or experiences, but what if that customer's needs have evolved? What if they're ready for something different, something they don't even know they want? Algorithms can't predict this; they can only replicate the past. By relying too heavily on AI-driven engagement strategies, brands risk reducing customers to a static profile, missing opportunities to surprise, delight, and connect on a deeper level.

Can AI Hear You?

It feels simple initially—AI certainly has audio recognition, speech-to-text capabilities, and sentiment analysis. It can "hear" us in a functional, even impressive way. But does it hear us in the way we, as humans, truly desire?

When we say we want to be heard, it's not a literal exercise in recording words or understanding the structure of a sentence. We're asking for something more profound: acknowledging our feelings, understanding unspoken needs, and connecting beyond surface-level communication. In relationships, being heard isn't about precision; it's about empathy, about feeling understood. It's that moment when someone truly "gets" you—a feeling that builds trust and comfort.

AI can listen, yes. But can it create those moments that leave someone feeling genuinely understood?

Imagine confiding in a friend about a difficult day. A friend who "hears" you at a basic level might say, "I understand you're frustrated." AI can simulate this response, using sentiment analysis to detect frustration in your tone or words. Advanced AI models can even adapt their responses to sound empathetic: "I'm sorry to hear that. It sounds like it's been a tough day." AI could adjust its tone, its timing, and its phrasing to make the interaction feel smoother.

AI can also recognize patterns and emotional cues. It can analyze millions of interactions to identify what phrases work best to calm an angry customer or reassure someone expressing sadness. This ability allows AI to triage issues in customer service, flag urgent emotional needs, or suggest personalized solutions at scale.

But think about the friend who doesn't just say, "I understand you're frustrated." Instead, they show up at your door with a pint of ice cream or invite you out for a drink to commiserate. Or consider the power of an 8-second hug—something Dr. Christy Kane (2024) identifies as a simple yet transformative act. A physical embrace releases oxytocin, reduces stress, and strengthens bonds, creating a connection words alone cannot achieve. It's the warmth, trust, and presence of these human moments that AI, for all its sophistication, cannot replicate.

While AI can craft empathetic responses and analyze emotional cues with remarkable accuracy, it cannot transcend analysis. No matter how refined its responses become, it won't offer a hug that reduces cortisol or hold space in a way that turns grief into gratitude. These moments live in the realm of human connection, beyond the reach of machines.

The Best-Case Scenario

If AI reaches a point where it can seamlessly replicate empathy in four key areas—recognizing emotions, tailoring responses, adapting to context, and maintaining immersion—it could revolutionize support systems. AI could provide emotional guidance to millions who feel isolated or overwhelmed, mainly when human intervention isn't accessible. For example:

- A caregiver might rely on AI to soothe a frustrated dementia patient, helping them navigate confusion with calm and personalized reassurance.
- Emergency response systems could deploy empathetic AI to comfort someone experiencing a crisis until human help arrives.

Even in this best-case scenario, the gap remains clear. Genuine empathy isn't just about *what* is said—it's about a more profound, nuanced understanding of shared life experiences. AI may say the right things, but it doesn't *mean* them. And when the realization dawns that the care offered is calculated, not heartfelt, something breaks.

Imagine that mid-conversation, the AI makes a mistake—an awkward response or a comment that feels out of place. Suddenly, the immersion shatters. You're reminded that you're not speaking with someone who understands you; you're interacting with a machine. It's like catching someone in a lie—trust immediately evaporates, and confusion or frustration creeps in. This fragility underscores the limits of simulated empathy. Once the illusion breaks, the emotional connection is gone.

But should we continue attempting to build empathetic AI? Perhaps not entirely. AI can manage the functional aspects of listening—analyzing data, spotting trends, and suggesting actionable insights—freeing humans to focus on the emotional core of conversations.

AI could assist customer service agents by summarizing a customer's frustration in real time or suggesting empathetic responses, empowering the agent to engage more deeply. In mental health, AI might monitor tone and sentiment to alert therapists when a client seems especially distressed, allowing for timely intervention.

However, for AI to be genuinely empathetic, it would need to recognize emotions and, more critically, also experience them—and that's the line it can't cross. Maybe that's okay. AI doesn't need to replace human empathy; it can amplify it. And perhaps that's where its most significant strength lies: building opportunities for humans to understand empathy better and, in my opinion, help build more empathetic companies.

Interpersonal Intelligence vs. Conversational AI

Interpersonal intelligence enables humans to engage deeply, build trust, and foster relationships through empathy, active listening, and non-verbal cues. Humans intuitively navigate sarcasm, humor, and emotions, adapting their communication to meet the unique needs of each interaction.

Conversational AI facilitates communication through natural language processing (NLP), handling routine tasks, and providing real-time support. However, it operates transactionally, lacking genuine understanding or emotional connection. It struggles with subtleties like sarcasm or body language, often delivering responses that feel mechanical. This gap limits its ability to replicate the relational depth of human conversations.

Interpersonal Intelligence and Conversational AI for Customer Engagement

- **Training AI on Human-Led Insights:** Leverage insights from human interactions to refine AI models, teaching AI to better interpret subtle cues like humor or sarcasm, while acknowledging it will never fully replace human intuition.

- **Integrate AI and Human Touch Seamlessly:** Ensure smooth handoffs from AI to human agents. When a customer's needs exceed what the AI can handle, train AI to know when to let go and make the transition should feel effortless without losing context. (A customer should NEVER need to repeat themselves about their issue.)
- **Measure Success Beyond Efficiency:** Don't just track metrics like resolution time or chatbot usage. Measure emotional engagement points, and human handoffs and interactions to gauge the true impact of your approach.

Chapter 8

FULFILLMENT IN FOCUS
Intrapersonal Intelligence in Customer Success

A good friend was asked to speak at a conference—a big opportunity for him but one that filled him with doubt. Each night leading up to the event, we prepared together, working through his key points, anticipating questions, and honing his delivery. We didn't focus on making it flawless; instead, I encouraged him to think about his central message and what he wanted the audience to take away. I reminded him to consider the listeners' perspective—what would resonate with them, what would serve them best. This process of understanding his own purpose and the impact he wanted to create wasn't just a practice exercise; it was a lesson in using intrapersonal intelligence, the kind that comes from looking within, recognizing one's own motivations, and aligning them with the needs of others.

On the day of the event, he stood in front of a room of fifty people. After days of rehearsal and revision, his delivery wasn't robotic or overly polished; it was real, it was him, and it connected deeply with his audience. They were engaged, nodding along, leaning forward. The positive energy in the room was undeniable. By the time he finished, the audience's enthusiasm and the conversations afterward made it clear: he hadn't just presented a series of ideas; he'd actually delivered something meaningful.

Reflecting on that experience, I realized it underscored the heart of customer success (CS). It's about being available to the customer, coaching them, and helping them realize their goals in ways that align with who they are and what they aspire to achieve. Just as I'd encouraged my friend to step into his role with authenticity and purpose, businesses need to foster a similar sense of alignment. Customer success means focusing on the aspirations and needs of the people you serve, not just pushing internal metrics.

This journey of preparation wasn't simply about "getting the job done." It was about building confidence and empathy, about the transformative power of intrapersonal intelligence—tuning into one's inner compass to understand what genuinely matters. And for businesses, tuning into this kind of intelligence means asking important questions: What will actually help our customers succeed? What do they truly care about, and how can we support them in achieving that?

The *2024 State of AI in Customer Success* report by the software firm Gainsight reveals that while more than 50% of CS organizations are now using AI, it hasn't yet reached "essential" status. Instead, AI remains a productivity tool, often adopted at the initiative of individual team members rather than through organizational mandates or structured programs.

In the end, it's not just about the service or the product; it's about showing up for customers the way you would for a friend, guiding them not only toward their goals but also to a place of confidence and clarity about their purpose. True customer success isn't measured by what you deliver but by the difference you make in the goals, lives, and journeys of the people you serve.

Real Success

Fifteen years ago, Bruce Temkin, as part of The Temkin Group, introduced a trio of customer experience (CX) metrics: success, effort, and emotion. Initially designed to measure the holistic customer experience, Temkin's metrics have retained their foundational simplicity and effectiveness today.

Decades earlier, in 2003, Fred Reichheld introduced the Net Promoter Score (NPS) in his seminal *Harvard Business Review* article, *The One Number You Need to Grow*. Developed in collaboration with Bain & Company and Satmetrix Systems, NPS was intended as a straightforward metric to gauge customer loyalty, asking the now-famous question: "How likely are you to recommend this product or service to a friend or colleague?" Its simplicity and predictive power quickly made it a popular tool for organizations worldwide.

Despite the reliability of Temkin's metrics, many organizations continue to embrace models like NPS to gauge customer sentiment. However, in today's fragmented, multidimensional environment, where consumers constantly transform, NPS is increasingly insufficient to capture customer interactions' nuances.

As Fred Reichheld himself acknowledges since developing NPS, "I had no idea how people would mess with the score to bend it, to make it serve their selfish objectives." This distortion reduces the accuracy of NPS as a KPI and, in turn, diminishes its value as a strategic tool for improving CX. While NPS remains a widely used benchmark, the evolving complexity of customer interactions calls for more nuanced approaches to understanding and enhancing the customer experience.

The ongoing focus on NPS, often more applicable to brand perception than service experience, has led some organizations to overlook the rich insights that the success, effort, and emotion framework can deliver, especially in better understanding exactly what customer experience is.

Understanding the Metrics: Success, Effort, Emotion

Temkin's model is straightforward:

- **Success:** The degree to which customers can accomplish their goals
- **Effort:** The difficulty or ease in accomplishing their goals
- **Emotion:** How the interaction makes customers feel
- Customer Experience = Success + Effort + Emotion

When these metrics are combined, they provide a comprehensive snapshot of how well a company is delivering on its promise of a positive customer experience. CX professionals who understand the importance of these metrics can build a clearer picture of customer satisfaction and loyalty, driving actionable insights to enhance service delivery.

But be aware that this process is about perspective. Too often, I see organizations with KPIs that are self-serving and profit-focused. Reversing your perspective on the critical nature of customer metrics is the key to tracking their success. Here's how:

Success: Do Your Customers Achieve Their Goals?

Success, in the context of customer experience, focuses on the tangible outcomes of interactions. Did the customer accomplish what they set out to do? Whether completing a purchase, resolving a service issue, or accessing information, this metric helps measure how well the experience is aligned with the customer's objectives.

Customers can have multiple experiences across multiple channels—online, mobile apps, in-person, AI, call centers—making the definition of success complex. The simplicity of this measure makes it universally applicable. By understanding the degree to which customers meet their goals, businesses can identify friction points in the customer journey and streamline processes.

When success is measured, it becomes easier to pinpoint bottlenecks and inefficiencies. The brand risks eroding trust if many customers fail to achieve their desired outcomes. Where customer expectations for efficiency are high, the success metric is about your customer's goals, not your ability to achieve your own goals or profits.

Success: What Is NOT a Customer Goal?

1. **Renewing a Subscription or Contract:** This is actually a business goal. While renewals are a crucial measure of customer retention, their existence reflects the company's desire to keep customers locked into a financial commitment rather than an indicator of whether they achieved their goals. A better customer success metric would be how the product or service helped the customer solve a problem or meet their objectives.

2. **How Quickly You Onboard a New Customer:** Speed of onboarding is often considered a success

metric for companies. Still, customers don't care how fast you onboard them; they care about being ready to use your product or service effectively. A better success metric would be whether the customer feels prepared and equipped to reach their goals after onboarding rather than how fast the process took.

3. **Uptake of New Features:** Businesses love tracking how many customers use newly released features, but the customer's goal is different. The customer's view of success is based on whether those features help them solve a problem or achieve their goals, not on how many features they've adopted (especially if you force them to use the feature.)

Effort: The Customer's Experience with Friction

While success measures the outcome, effort gauges the journey. How easy or difficult is it for customers to accomplish their goals? This metric is particularly relevant in customer service, where ease of resolution is critical. Research consistently shows that customers are likely to remain loyal to a brand when their interactions are simple and friction-free.

Reducing customer effort has become an increasing focus in CX. In today's omnichannel environments, customers expect seamless transitions between different touchpoints—whether they start on a mobile app and finish on a website or interact with customer support via chat and later over the phone. High effort leads to frustration, while low effort correlates with customer satisfaction and loyalty.

A recent survey by Gartner found that 96% of customers who experience high-effort service interactions become disloyal compared to just 9% who have low-effort interactions. This demonstrates that effort is not just about convenience but also a critical driver of loyalty.

Effort: What Is NOT a Customer Goal?

1. **Time Spent Resolving Issues:** Many businesses measure success by reducing the time it takes to resolve a customer's issue, but speed alone does not capture the ease of the experience. Customers might appreciate a quick resolution but value a seamless, stress-free interaction. Cutting down resolution time doesn't necessarily mean the process was easy or pleasant.

2. **Reducing Call Transfers:** Businesses often pride themselves on metrics like "fewer transfers between departments," but for customers, the real goal is to solve their problem during one interaction. Whether the situation involved one person or several isn't as important as resolving the issue efficiently. Efforts should measure whether the customer felt they had to jump through hoops to solve their problem, not how well your internal routing systems work.

3. **Improved Internal Efficiency:** Companies frequently track how streamlined their processes have become, aiming for lower overhead or faster workflows. While internal efficiency may benefit the company, it's not the customer's goal. The customer cares about how easy and intuitive the experience is for them, regardless of how much time or money the company saved on the back end.

Emotion: How Does it Make Customers Feel?

Emotion might be the hardest of the three metrics to quantify, but it is arguably the most critical. How a customer feels during and after an interaction plays a pivotal role in shaping brand perception . Positive emotions like happiness or relief build loyalty, while negative emotions like frustration or indifference can drive customers away.

The psychological aspect of customer experience is often missed. It's about understanding how the interaction makes customers feel, not just about the outcomes they achieve or the effort they exert. While brands have become adept at measuring tangible outcomes, they are often less successful at gauging emotional reactions—even though these reactions heavily influence future behavior.

Emotional insights into customer experience programs can differentiate a brand from its competitors. Emotionally engaged consumers are likely to become repeat customers, promote the brand, and have a higher lifetime value. Businesses can tailor their strategies by measuring emotion to evoke positive feelings while creating deeper customer connections.

Emotion: What Is NOT a Customer Goal?
1. **Customer Satisfaction Scores (CSAT):** While relying on CSAT surveys to measure emotional impact is tempting, these scores often reflect generic satisfaction rather than the depth of the customer's emotional connection to the brand. Customers don't set out to feel satisfied after an interaction; they want to feel valued, respected, and supported. Satisfaction scores are helpful but don't necessarily equate to emotional success.

2. **Minimizing Negative Feedback:** Businesses sometimes focus on reducing the volume of negative reviews or complaints, assuming this indicates positive emotions. However, lowering complaints doesn't mean customers are emotionally connected or loyal; it might just mean they didn't bother to voice concerns. The goal should be cultivating authentic, positive emotions during interactions rather than just reducing negativity. It's also important to recognize that most people don't believe reviews are genuine (positive or negative.)

3. **Promoter Scores (NPS):** While NPS can help gauge overall brand loyalty, it doesn't directly reflect the emotional experience tied to individual interactions. Customers' emotional goals are tied to feeling understood, trusted, and appreciated during each touchpoint, not simply whether they'd recommend the brand to others. NPS reflects a future behavior (referral), but it doesn't capture the emotional depth of the current interaction.

How AI Can Undermine Customer Success

AI in CS promises faster, more personalized experiences that anticipate customer needs. Claiming that AI can foresee and fulfill a customer's needs even before they articulate them is exciting. In reality, however, there's a risk that AI will focus too much on automation at the expense of genuine human connection. CS requires more than an efficient algorithm; it demands empathy, understanding, and a commitment to helping customers achieve their unique goals.

Laura Lakhwara, senior manager of customer success at UiPath, highlights AI's strengths in this space:

"AI has transformed the way we understand our customers' goals. What used to take hours—gathering data through sentiment analysis, mining interactions, and scouring social channels—can now be streamlined and turned into actionable insights. AI doesn't just collect and organize customer data; it empowers teams to create collaborative dashboards, automate responses, and guide customers to their next milestones. By leveraging AI, CS teams gain a nuanced view into individual customer health and broader industry trends, enabling us to serve customers with precision and foresight."

While Lakhwara emphasizes AI's role in organizing and analyzing customer data, others caution against relying solely on technology. Denise Stokowski, SVP of product management at Gainsight, articulates this balance well:

"Despite its technological roots, AI is a path to more human-centric interactions. So we're working to ensure that AI enhances rather than replaces human connections. Because at the end of the day, humans are still best at navigating nuanced conversations. You need that skill to truly get the most value. AI is going to radically make customers and customer success better."

In Stokowski's view, AI should act as a force multiplier, not a substitute, enhancing the role of customer success specialists rather than diminishing their human connection.

Research from software company Master of Code (2024) supports the optimism surrounding AI in customer service. In companies already using AI in their support units, 91% report satisfaction with the results, and 56% feel more optimistic about AI's impact on success compared to last year. The data shows key benefits: improved service (69%), decreased wait times (55%), and streamlined workflows (54%). In practical terms, 94% of conversational AI and chatbot users report increased productivity, 92% note faster issue resolution, and 87% see reduced agent effort. AI's efficiencies don't just improve customer satisfaction (48%); they also lead to better data utilization (41%) and significant cost savings. Of those surveyed, 65% report lowered expenses, and 53% cite reduced operational costs in AI-driven call centers.

These numbers show that when AI and customer success teams align, AI can remove friction, streamline processes, and empower representatives to engage with customers in more impactful ways. However, AI remains only one piece of the equation. Lakhwara and Stokowski agree that human-centered service is essential for building trust, handling complex issues, and ensuring customers feel valued. Ultimately, AI's potential lies in supporting a more human approach to customer success, freeing up specialists to focus on the nuanced, relationship-driven aspects that matter most in a customer's journey.

In the end, AI's real value is in its ability to help customer success teams deepen connections rather than replace them. As businesses increasingly adopt AI, leaders must remember that technology can improve efficiency, but only fellow humans can make customers feel truly understood.

The Dance Between Automation and Humanity

Let's not confuse convenience with connection. While sentiment analysis chatbots, predictive analytics, and natural language processing (NLP) tools streamline tasks and improve efficiency, they fall short of replicating the depth of understanding that actual customer success demands. This gap highlights the delicate balance between automation and humanity—and why human intrapersonal intelligence remains an irreplaceable part of the equation.

Narrow AI systems, such as chatbots and speech analytics, have revolutionized the customer success landscape. These tools process vast amounts of data, detect sentiment in a customer's tone, and predict behavior patterns based on historical interactions. For example, sentiment analysis AI can identify frustration in a customer's voice during a phone call and automatically escalate the issue to a human representative trained to handle complex emotions. Predictive analytics can anticipate a customer's needs before they articulate them, enabling proactive solutions.

But what happens when the conversation isn't about "What time do you open?" or "Where's my order?" but rather "I feel let down," or "I need help, but I don't know how to ask for it?" This is where the limits of Narrow AI become glaringly obvious.

Sentiment analysis tools, for instance, are programmed to detect emotional cues like frustration, happiness, or anger based on data patterns such as tone, pitch, or word choice. Yet emotions are far more nuanced than a soundwave or a string of text can convey. A frustrated customer might mask their irritation with humor, while an anxious caller might remain calm on the surface but hesitate to share their real concerns. AI systems often fail to read between the lines, missing the subtle signals that human beings can instinctively understand.

Predictive analytics face similar challenges. While these tools forecast a customer's next move based on historical trends, customers are, in fact, not static patterns—they are individuals with evolving needs, shifting priorities, and emotions that often defy past behavior. When a customer's actions deviate from their historical data, predictive tools can produce automated responses that feel tone-deaf or irrelevant, leading to customer alienation.

And then there's context, the intangible layer of meaning that AI cannot grasp. It doesn't understand if a customer calls about a delayed order on the day they received disappointing news at work. It doesn't see the personal significance of the product or service in the customer's life. Humans, with their intrapersonal intelligence and ability to empathize, do.

In the dance between automation and humanity, AI excels at efficiency and scale, but human understanding adds the depth and authenticity customers crave. Together, these elements must work in harmony to achieve true customer success.

Intrapersonal Intelligence: The Missing Ingredient

Customer success is often defined as helping clients achieve their goals, but its foundation lies in self-awareness. Gardner (2011) defined intrapersonal intelligence as:

> the capacity to understand oneself and one's thoughts and feelings, and to use such knowledge in planning and directioning one's life. Intrapersonal intelligence involves not only an appreciation of the self, but also of the human condition. It is evident in psychologists, spiritual leaders, and philosophers.

While customer success focuses outwardly on the client's experience, it begins with the professional's inward clarity.

Steve Jobs exemplified high intrapersonal intelligence through his deep understanding of his own vision, motivations, and values, which allowed him to focus relentlessly on creating products that aligned with his passion for innovation and design. His ability to reflect on his personal experiences—such as being ousted from Apple and returning with greater clarity—showcased his capacity for growth and emotional regulation. Jobs' introspection and self-awareness enabled him to communicate his ideas authentically and inspire others to embrace his vision, leaving a lasting impact on technology and culture.

Self-aware professionals excel in customer success because they can better engage with customers empathetically, understand their motivations, and manage their personal emotions in the process. This self-awareness enables thoughtful responses, genuine listening, and the ability to connect deeply with customers. A support agent handling a frustrated caller might recognize the customer's tension and stress and use that awareness to approach the interaction calmly. American poet and civil rights activist Maya Angelou once said, "People will forget what you said, people will forget what you did, but people will never forget how you made them feel." These individuals transform a challenging moment into trust and connection by offering reassurance or a relatable story. These human nuances are beyond AI's capability.

AI tools like sentiment analysis and predictive analytics are exceptional at processing data and detecting emotional cues. They can identify frustration in a customer's tone and escalate issues efficiently. However, as futurist and

AI expert Bernard Marr (2024) points out, "It doesn't really have new ideas in the same way that humans do. Its creativity is only informed by data rather than by feelings, emotions, original thoughts, and personal experience of the world." AI cannot understand emotions, the context of an interaction, or the personal experiences that inform human responses. It can assist with routine tasks but cannot replicate the depth and adaptability of human connection.

Intrapersonal intelligence elevates customer success from transactional problem-solving to building meaningful relationships. By starting with self-awareness, professionals create personalized, impactful interactions that inspire loyalty and trust. This human touch fills the gaps AI cannot bridge, ensuring customer success teams are not just meeting expectations but also exceeding them in ways that resonate deeply with clients. While Tidio, an AI support desk company, predicts that 95% of customer interactions will be powered by AI 2025, it is clear that technology alone cannot replicate the empathy and connection required to build lasting relationships. A PwC (2024) survey underscores this gap, revealing that 59% of consumers feel companies have lost touch with the human element of customer experience. These statistics highlight the irreplaceable value of human-driven interactions, where intrapersonal intelligence creates the foundation for meaningful engagement.

Customer success requires a mastery of adaptability. Clients vary in needs, communication styles, and expectations, and each interaction demands flexibility. But adaptability starts with self-awareness. If a customer success professional understands how they typically respond to different situations, they can adjust their behavior, shift their communication style, and put clients' needs first, without letting personal tendencies or preferences interfere. By blending adaptability with self-awareness, professionals can address clients' unique circumstances in ways that AI-driven systems cannot, bridging the gap between efficiency and empathy.

The Automation Trap

One of the dangers of relying too heavily on Narrow AI is falling into the trap of transactional engagement. Chatbots can answer questions, and Interactive Voice Response systems (IVR) can route calls, but these interactions often feel cold and impersonal. Customers leave the conversation feeling like just another data point rather than an individual whose needs and concerns matter.

Transformational engagement, on the other hand, goes beyond answering the "what" and addresses the "why." It's about understanding the emotional context of a customer's experience and responding in ways that make them feel heard and valued. Achieving this requires the kind of introspection and empathy that only humans possess.

That's not to say AI doesn't have a role in customer success—it absolutely does. But its role should be to enhance human interactions, not replace them. AI can handle the routine and mundane, freeing human agents to focus on the complex, emotional, and nuanced conversations that truly define customer success.

Empathy Begins with Self-Awareness

The best customer success professionals don't just listen to their clients; they also understand them on a deeper level. But empathy isn't a skill that exists in a vacuum. It's cultivated through a deep self of self. When professionals know what triggers their own emotions, they're better positioned to recognize similar emotional signals in others.

Think of intrapersonal intelligence as a tool for empathy training. By managing their own emotional landscape, customer success professionals create a stable foundation that allows them to genuinely connect with clients' challenges, fears, and goals. In a customer success role, this level of empathy can mean the difference between a client feeling truly supported and one feeling like they're just another number on the dashboard.

Empathy shaped by self-awareness is less reactive and more attuned. When we are grounded, we're not distracted by our own internal dialogue; we're fully present and able to interpret the client's experience without projecting our own emotions or assumptions. The best customer success professionals use intrapersonal intelligence to build this kind of balanced empathy, a powerful way to make clients feel seen and understood.

One Customer Success Team

Customer success is more than a scoreboard of metrics; it's a philosophy that stands or falls on one question: *Did the customer achieve what they set out to do?*

Metrics like Customer Satisfaction Score (CSAT), Net Promoter Score (NPS), or even Customer Lifetime Value (CLV) are the staples of today's dashboards. But here's the uncomfortable truth: they measure our success, not the customer's. CSAT reflects whether someone liked the interaction, but it doesn't guarantee they got what they needed. NPS, which has become a near-obsessive metric, merely hints at loyalty and can't tell us if we solved the customer's problem. Even CLV, a measure of how much we can profit from a customer, is ultimately an exercise in evaluating our own bottom line.

We've confused measuring for succeeding.

Customer success should mean one thing: the customer succeeded in their goals with our help. Anything less, and we're just keeping score for ourselves. Instead of asking how many customers renewed, we should ask how many achieved their KPIs. Imagine a new metric, a Customer KPI Score, CKPI, that tells us how often we helped customers meet their own goals. It's a metric that forces us to keep fulfillment in focus, holding ourselves accountable not for revenue but for results.

Consider this: if a customer purchased a tool from us to streamline their workflow, customer success should be about whether that workflow became more efficient. If they signed up for our service to reduce costs, success means not just that they're satisfied with the interface but that they actually saved money. The goal is alignment, not applause.

Executives love KPIs because they can be manipulated to show ROI. AI-driven analytics only increase this illusion. Machine learning might be able to process data at lightning speed, but it cannot define success. Algorithms serve the inputs we choose, and all too often we choose inputs that show what we want to see. The result? Businesses "prove" success while customers remain stuck, often unable to achieve the goals they came to us for in the first place.

Real customer success doesn't happen because a data model tells us we're doing well; it happens when there's an honest reckoning, a moment of introspection about what we're actually measuring and why. Metrics should be mirrors, not trophies. Are we genuinely invested in the outcomes of those we serve, or are we placating ourselves with scores that only reflect our side of the story?

And here's the larger shift: customer success should be the thread that unites service, support, and the success team into one cohesive whole. These three S's should be on the same page, committed to one singular philosophy: *our success is only as real as theirs.* Support shouldn't just mean handling a customer complaint; it should mean helping the customer clear the hurdle in their way. Service shouldn't only mean satisfying the inquiry; it should mean offering an experience that leaves them more capable than when they arrived. Success shouldn't just be your metric for renewal or churn; *it should mean we moved the needle on their goals.*

When we see customer success as the pinnacle of everything we do, our teams stop competing for metrics and start collaborating for outcomes. It's no longer about satisfying or closing. It's about achieving and empowering.

If we want to thrive in the long term, we must shift our gaze outward, toward the horizon of our customer's ambitions. The metrics we've relied on aren't inherently bad, but they're incomplete. As we expand our view, we need metrics that keep us honest, that push us to ask hard questions: *Did the customer actually succeed? Did we play a role in that success?* When our success metrics reflect the actual journey of those we serve, we're no longer counting check marks on a dashboard; we're creating impact.

Success for the customer is success for us. That's the essence of customer success, not as a department, but as a calling.

Intrapersonal Intelligence vs. Sentiment AI

Intrapersonal intelligence involves profound self-awareness, allowing individuals to reflect on emotions, values, and motivations to grow personally and professionally. This introspection fosters resilience, adaptability, and deeper self-understanding.

Sentiment AI analyzes tone, text, and behavior data to identify emotions and provide insights. While it can efficiently detect sentiment and adapt processes, it lacks accurate self-awareness or introspection. Without a "self" to reflect on or learn from, Sentiment AI cannot replicate the depth of human emotional understanding or the personal growth enabled by intrapersonal intelligence.

Intrapersonal Intelligence and Sentiment AI for Customer Success

- **Self-Aware Leadership:** Encourage professionals to pair Sentiment AI insights with their own self-reflection, enabling them to approach customer interactions with both data-driven context and authentic emotional intelligence.
- **Dynamic Feedback Loops:** Employ Sentiment AI to monitor patterns in customer sentiment and provide actionable feedback, while humans interpret these patterns to make strategic adjustments with empathy and foresight.
- **Proactive Customer Care:** Combine Sentiment AI's ability to detect dissatisfaction early with human problem-solving skills to address issues before they escalate, strengthening customer trust and loyalty.

Chapter 9

ETHICAL BLIND SPOTS
Moral Intelligence in Customer Loyalty

AI has significantly advanced the personalization and efficiency of customer interactions, but these improvements alone have not automatically increased brand loyalty. Consumers seek more than seamless transactions; they value trust, authenticity, and meaningful connections with the brands they follow and the communities in which they participate. Building loyalty requires moving beyond efficiency to foster relationships grounded in shared values and emotional connections, rather than purely transactional exchanges.

Ethical AI plays a key role in building this trust. This term refers to the development of AI systems that uphold moral principles, ensuring fairness, transparency, and accountability in their operations. The goal is to create technology that benefits society while minimizing harm. However, challenges like algorithmic bias can lead to unfair outcomes and reinforce societal inequalities. A 2020 report from *The Enterprisers Project* revealed that:

45% of consumers who experienced negative interactions with AI would share their dissatisfaction with
others, discouraging engagement with the organization. Additionally, 39% would switch to higher-cost
human channels, and 27% would reduce trust or stop
engaging with the organization altogether.

Supporting this perspective, a 2022 McKinsey report
revealed that organizations adept at building digital trust
are more likely to see annual growth rates of at least 10%
on their top and bottom lines. Prioritizing Ethical AI is
critical for building trust, mitigating legal and reputational
risks, and ensuring AI systems serve society equitably. As
AI becomes increasingly integrated into everyday life, ethical considerations must guide its development and deployment to prevent unintended consequences.

However, Ethical AI is not directly related to data
security. Instead, it is focused on internal accountability
within an organization. While safeguarding consumer data
is essential, Ethical AI (or also referred to as Responsible
AI) focuses on ensuring fairness, transparency, and moral
responsibility in AI systems' development and deployment.
It requires organizations to actively reflect on their values
and decisions, asking, "What is the right way to use this
technology for the benefit of all?"

This approach establishes a foundation for trust. Consumers recognize and appreciate transparency in how AI
operates and the principles guiding its application. Ethical AI may not eliminate every mishap, but it fosters an
environment where companies can openly communicate
their ethical frameworks and address mistakes with integrity. Such honesty resonates with consumers, who increasingly value accountability over perfection. Acknowledging

potential biases in algorithms or unintended outcomes demonstrates a commitment to continuous improvement, reinforcing trust even in challenging situations.

Ethical AI also elevates the moral intelligence of an organization, creating alignment between its internal decision-making and external brand promise. This alignment strengthens the perception of authenticity, which is critical for building trust and fostering loyalty. Consumers want to support companies that prioritize ethical practices, even when these practices do not directly involve data protection. By embracing Ethical AI as an internal tool for moral accountability, organizations send a powerful message: trust is not just a marketing strategy; it's actually a core value.

Ultimately, trust catalyzes loyalty, and Ethical AI provides the transparency necessary to nurture that trust. It shifts the conversation from transactional exchanges to meaningful, value-based relationships. In doing so, Ethical AI becomes more than a framework for technological development; it becomes a cornerstone of consumer confidence and enduring loyalty.

The Trust Factor

Trust remains central to brand loyalty. A 2024 survey by Linearity, a digital design company, revealed that only 23% believe brand consistency can boost revenue, yet 81% of consumers say they need to trust a brand before purchasing. This trust develops through transparent, authentic interactions that AI alone struggles to replicate. 94% of consumers show loyalty to brands that prioritize transparency.

Deloitte's research shows that on average, nearly two-thirds of U.S. consumers belong to between one and five loyalty programs, but they use less than half of them. For retailers, this suggests a need to develop engaging programs that drive actual participation and foster true loyalty. Another study found that loyalty program members reported a 61% higher level of trust in their brand compared to non-members.

Ulta Beauty is a prime example of leveraging loyalty programs effectively. A 2024 Deloitte survey found that Ulta grew active members from 30.7 million in 2020 to 40.2 million in 2022. More than 94% of its annual sales came from these members, prompting Ulta to update its loyalty program in January 2024 with "additional perks and a more personalized birthday benefit."

Deloitte (2024) also found that retail executives are prioritizing AI-driven product recommendations, yet only half feel confident in their company's capability to build AI tools across the business. However, consumers increasingly prefer brands that align with their values and engage in community-building, over new tools such as AI.

CapitalOne's research (2024) also shows that 65% of retail revenue comes from loyal customers, who spend 67% more on average. Globally, 74% of consumers claim brand loyalty, and 75% would switch brands for better loyalty rewards. Across all age groups, trustworthiness and transparency matter. Sixty percent of consumers rated these as the top brand traits in 2022, up from 55% the previous year. Meanwhile, 56% report stronger loyalty to brands they feel understand them. This underscores the need for brands to engage actively and demonstrate commitments to shared causes, values, and community.

AI can enhance experiences, but can it foster trust? While terms like "Ethical" and "Responsible" are becoming common, skepticism remains. A 2024 study by the creative consultancy Lippincott found that less than a third of consumers feel AI meets their expectations, with 40% expressing doubts. Surprisingly, trust in AI doesn't vary significantly by age; 24% of 18- to 24-year-olds trust brands to use AI responsibly, compared to 18% of consumers aged 65 and older.

Chris Ciompi, senior partner at Lippincott, notes, "Brands must find authenticity in the artificial to earn consumer trust." Yet, a Salesforce study indicates declining trust across the board, with 72% of consumers reporting less trust in companies than a year ago (Wood, 2024). Consumers often expect that AI will enhance their experience seamlessly, yet many are disappointed when brands fall short of this expectation.

Moral Intelligence over Ethical AI

Moral intelligence refers to the ability to apply principles of ethics, responsibility, compassion, and integrity in decision-making processes. Unlike traditional forms of intelligence, which emphasize logic, problem-solving, and analysis, moral intelligence is rooted in values and purpose-driven leadership. This makes it a vital component in building trust within an organization and fostering positive relationships with the broader community.

For businesses, moral intelligence entails acting ethically not simply to meet compliance requirements but also because it reflects a commitment to doing what is right. Decisions rooted in moral intelligence prioritize the welfare of employees, customers, society, and the environment, aligning organizational actions with core values and purpose. Psychologists Doug Lennick and Fred Kiel, who developed the concept of moral intelligence in 2005, emphasize that individuals with high moral intelligence demonstrate a keen sense of right and wrong, enabling ethical decisions even in complex or ambiguous situations. Such integrity strengthens trust, credibility, and loyalty—qualities that are crucial as consumers increasingly scrutinize corporate motives.

Ethics has similarly become a defining challenge in the realm of artificial intelligence as businesses grapple with the societal implications of generative AI. A 2024 IBM Institute for Business Value survey reveals that 56% of organizations are delaying significant AI investments due to unclear standards and regulations, while 72% are prepared to sacrifice potential benefits over ethical concerns. Phaedra Boinodiris, global leader for trustworthy AI at IBM Consulting, highlights that "creating ethical AI is not strictly a technical problem but also a sociotechnical problem." Building responsible AI systems requires the expertise of interdisciplinary teams—including linguists, philosophers, and individuals with varied life experiences—to address critical questions such as, "What are the unintended effects of AI?" and "Is this AI solving the problem we need it to?" Diversity within these teams ensures models are not only responsibly curated but also more precise and effective.

The importance of AI ethics has gained traction as a competitive differentiator. According to IBM (2024), 75% of executives recognize AI ethics as a key driver of differentiation, while 54% view it as strategically vital, (McKendrick, 2024). Ethical AI practices yield multifaceted returns, spanning economic benefits such as cost savings, long-term infrastructure advancements, and enhanced brand reputation with stakeholders. Boinodiris advocates for ongoing education within leadership, urging companies to "engage your savviest AI ethics experts" to help executives reframe AI ethics as a source of innovation and growth. This holistic approach fosters forward-thinking strategies that align technological progress with societal well-being, ensuring AI systems serve humanity responsibly and effectively.

Why Moral Intelligence Matters in Business

Moral intelligence serves as a balancing force that prioritizes human values over pure efficiency. There are three main reasons why moral intelligence is particularly crucial in every aspect of business:

1. **Trust-Building:** Trust is fundamental to long-term success, but it cannot be generated solely through AI. While AI can predict behaviors and personalize experiences, it lacks the emotional understanding and ethical grounding required to build trust on a deeper level. Consumers want to feel that the companies they support are trustworthy and value-driven, making moral intelligence essential to achieving genuine connections with customers.

2. **Employee Engagement and Retention:** Moral intelligence in leadership contributes to a positive work culture. When companies act responsibly and ethically, they foster an environment where employees feel valued and secure. According to a Deloitte (2024) survey, 73% of millennials said they are more loyal to employers who prioritize ethics, and nearly 60% of employees reported they would stay longer with companies that prioritize social and environmental impact. Moral intelligence, therefore, not only enhances external trust but also strengthens internal commitment and loyalty.

3. **Navigating Ethical Complexities in AI:** As companies adopt AI for decision-making, they face

ethical questions about bias, privacy, transparency, and accountability. AI algorithms may inadvertently reinforce stereotypes or create unfair outcomes, but moral intelligence can help guide responsible AI use. Leaders with strong moral intelligence are more likely to recognize these ethical complexities and take action to prevent harm.

Moral Intelligence to Build Trust and Community

While AI offers valuable tools for efficiency and insights, it lacks the moral framework that human-driven intelligence can bring. Here's how companies can leverage moral intelligence over AI to build trust and deepen their connection with customers and communities:

1. **Prioritize Transparency and Accountability:** One of the cornerstones of moral intelligence is honesty. Companies can foster trust by being transparent about how they collect, use, and store customer data. Implementing clear policies and openly communicating these policies shows respect for consumer privacy, an area where AI-driven decisions are often scrutinized. A company might choose to disclose the criteria its AI algorithms use for personalization, ensuring that customers understand and feel comfortable with how their data impacts their experience.

2. **Embed Ethical Standards in AI Practices:** Moral intelligence can play a pivotal role in the development and deployment of AI systems. By embedding ethical standards into AI projects, companies can ensure that their technology respects human rights and equality. This might

involve conducting regular audits for bias in AI algorithms or establishing an ethics committee to oversee AI initiatives. A company might also adopt principles for Ethical AI, making fairness, inclusivity, and transparency guiding factors in their technology development. Companies like Google and Microsoft have already set up AI ethics boards to manage the social impact of their algorithms. By implementing similar measures, organizations demonstrate a commitment to ethics and can differentiate themselves as trustworthy leaders in technology.

3. **Empower Employees to Voice Ethical Concerns:** Creating an ethical workplace environment involves empowering employees to raise ethical concerns. Companies should foster a culture where employees feel comfortable discussing moral or ethical questions related to AI and other business practices. When employees can speak up about potential issues, organizations are better positioned to address problems proactively.

4. **Engage with Community and Social Causes:** Moral intelligence emphasizes not just profit but also societal impact. Companies that invest in social and environmental causes, particularly those aligned with their brand values, build trust within the broader community. This can involve supporting local initiatives, promoting environmental sustainability, or engaging in fair labor practices.

Companies that lead with moral intelligence will not only navigate the ethical challenges of AI more successfully but will also establish themselves as authentic, trustworthy brands that consumers respect and support. By fostering transparency, embedding ethical standards in AI practices, empowering employees, and engaging in meaningful social causes, organizations can use moral intelligence to build genuine connections that set them apart in a technology-driven world.

The Loyalty Effect: Lessons from a Comic Book Shop
Customer loyalty is often seen as the result of promotions, reward points, and exclusive discounts, but true loyalty is actually rooted in something more profound: meaningful relationships and trust. A small comic book store next to the theater where I worked in college showed me what customer loyalty can look like when it's built on personal connection.

The comic book shop was more than just a place to buy comics; it was a place of genuine familiarity and camaraderie. I visited there regularly, drawn not only by my favorite character, Superman, but also by the warm, welcoming experience the owner created. He didn't just sell me comics. He remembered my favorite series, kept up with the latest issues I was reading, and even went out of his way to suggest new comics that matched my interests. His recommendations were always spot-on, and his excitement for new releases was contagious. I never felt like just another customer; he made me feel like I was a valued part of the store's community. It was a relationship built not on transactions but on shared enthusiasm.

My loyalty to Superman has been lifelong, starting from when I was a kid. Superman's story has always fascinated me, not only for the superpowers but for his sense of justice and moral integrity. Recently, I've been rewatching *Lois and Clark: The New Adventures of Superman* with Dean

Cain and Teri Hatcher, and it's been a fun flashback to the '90s. Watching it now, it's amusing to see the pre-smart-phone world, but it's also remarkable to see the guest stars who appeared in its four-season run. Legends like James Earl Jones, Raquel Welch, Drew Carey, and even Adam West brought a different energy to the show, making it a true time capsule of the era.

Yet, what stands out most about *Lois and Clark* is the portrayal of Superman's unwavering moral compass. Un-like many modern characters, Superman has remained relatively untarnished as a brand throughout his 80-plus years. His commitment to justice and the betterment of humanity is unwavering. Amidst an era where many heroes are flawed or ambiguous, Superman has stayed true to his values. For businesses, this kind of brand integrity is rare but valuable; it's a reputation that withstands the test of time. Superman's brand loyalty isn't just about nostalgia or powers; it's about the trust fans place in a character who consistently represents hope and courage.

In the same way, the comic book shop next to the theater wasn't just about the comics, it was about the re-lationship the owner fostered. The loyalty I felt wasn't driv-en by rewards programs or promotions; it was driven by a personal touch. That owner, like Superman, represented trustworthiness and authenticity. He knew his customers, shared their interests, and created a welcoming space where we felt valued.

Today, businesses invest millions in creating custom-er loyalty programs, but many overlook the foundational element of genuine connection. Customers aren't simply looking for discounts or points; they want to feel valued and understood. Big brands could learn from the humble comic book store owner: the key to loyalty is knowing your customers well enough that they feel seen, appreciated, and part of something meaningful.

Moving Beyond Transactions
If your conversations, digital strategies, experiences, and touchpoints don't include areas outside your business, then you view your customer solely as a transaction. This approach is not only limiting but also counterproductive. We have all heard the saying and seen the stats. People don't like to be treated as numbers. "The 84s" are a fantastic set of stats to remember.

- 84% of consumers say they buy from brands they feel emotionally connected to (Sprout Social, 2024).
- 84% of consumers are more likely to buy from a company that provides a personalized experience (Dynata, 2024).
- 84% of people said that "being treated like a person, not a number, is very important to winning their business" (Salesforce, 2024).

We must look beyond traditional journey mapping to truly understand and connect with customers emotionally and dynamically. This perspective encourages the creation of ecosystems that connect people not only to your business but also to other companies that intersect with their life journeys. By doing so, experiences become more emotional and interconnected and adapt as the customer evolves.

The Changing Dynamics of Trust and Loyalty
According to a 2022 McKinsey survey by Boehm et

al., trust in AI-powered products and data transparency has emerged as a critical factor shaping consumer decisions. Organizations capable of demonstrating Ethical AI practices and robust data protection measures meet expectations, foster loyalty, and drive growth. A 2022 McKinsey survey revealed that 70% of consumers believe the companies they patronize protect their data, indicating high expectations for digital trust. However, trust remains fragile; 57% of executives reported at least one material data breach in the past three years, creating a stark dissonance between consumer confidence and organizational performance. This misalignment risks businesses that fail to meet trust-related expectations as consumers increasingly scrutinize data practices.

Consumers are making trustworthiness a core aspect of their purchasing decisions however trust in AI remains complex, balancing optimism with caution. While a 2022 Capgemini report found that 49% of consumers trust their daily AI interactions, the opacity of AI processes often undermines confidence. According to technical writer Scott Clark (2022), the lack of explainability in AI, where decisions operate in a "black box," raises concerns about biases and accountability. Forrester and InRule (2022) supports this, with 58% of respondents worried about AI biases leading to inaccurate decisions. Transparency through Explainable AI (XAI) helps address these fears, as Vivun CEO Matt Darrow notes, by explaining the "why" behind predictions, fostering understanding and trust.

Transparency around AI and data policies has become a decisive factor for many consumers. Nearly half of respondents in the survey reported that they frequently consider switching to another brand if a company is unclear about how it uses its data. For younger demographics, like

Gen Z, this figure rises further. This shift demonstrates a growing demand for businesses to offer clear, accessible information about AI applications and data protection.

Ethical AI practices like cost and delivery time are becoming as significant as traditional purchase drivers. McKinsey's findings show that consumers value trustworthiness and data protection almost as much as price and delivery speed. More than half of respondents indicated they often or always buy products only from companies with a reputation for protecting consumer data. This trend underscores how trust, once considered ancillary, has become a central factor in fostering brand loyalty.

As AI becomes more integral to business operations, consumer behavior reveals an apparent demand for companies to prioritize Ethical AI practices. Transparency, accountability, and proactive risk mitigation are no longer optional; they are essential for maintaining trust and loyalty in a competitive marketplace. Businesses that rise to this challenge stand to benefit from consumer confidence and sustained growth and differentiation in the digital age.

The Psychology of Loyalty Programs

Loyalty programs are powerful tools for stimulating repeat behavior. From point-based rewards to tiered memberships, loyalty programs are designed to create a cycle of return visits, purchases, and interactions. By using incentives, companies can encourage customers to keep coming back. These programs often rely on behavioral triggers rather than emotional connections. The psychology behind loyalty programs is rooted in creating habits, not relationships, which can lead to loyalty that's transactional rather than genuine.

Take Starbucks, for example. Its loyalty program has been incredibly successful, and not because it builds deep connections but because it taps into the psychology of rewards. By offering stars for each purchase, which customers can then redeem for free items, Starbucks turns each visit into a chance to earn. But is this loyalty, or is it a habit? Many customers return not out of any strong brand connection but because the rewards structure encourages them to do so. They're engaged, yes, but at a level that's more about incentives than attachment.

When loyalty prioritizes transactions, brands risk losing customers to competitors offering better deals. Genuine loyalty, by contrast, builds trust over transactions. Customers return because they feel valued, aligned with the brand's purpose, or emotionally connected. This deeper loyalty—rooted in shared values rather than incentives—is far more resilient to competition and fosters lasting relationships.

When Technology Fails to Engage

While AI can analyze data and automate responses, it can't feel or truly understand human emotions. This lack of Emotional Intelligence is one of the biggest drawbacks of relying on AI to build loyalty. When customers reach out to a brand, especially with concerns or frustrations, they expect empathy. Yet, AI often misses the mark, providing canned responses that lack the nuance and empathy needed to reassure a customer.

Consider a customer reaching out to resolve a billing issue. An AI chatbot might quickly pull up information, but if it can't sense the customer's frustration or urgency, its responses can feel mechanical and cold. The customer's experience may improve in terms of efficiency, but it will likely suffer in terms of emotional connection. AI's inability to gauge emotional cues can lead to an experience that feels superficial and transactional, leaving customers feeling unheard.

Building loyalty requires more than just answering questions or solving problems; it requires understanding. When AI fails to address the emotional needs of customers, the connection weakens, and the loyalty it tries to build becomes fragile. Without emotional engagement, loyalty becomes a hollow concept, simulated rather than stimulated, based on convenience rather than trust.

The Future of Loyalty

As AI transforms customer engagement, businesses must decide whether to build simulated loyalty through convenience and incentives or foster genuine loyalty by cultivating meaningful connections. While simulated loyalty may be easy to engineer, it remains fragile, often dissolving with a better offer or a minor inconvenience. Genuine loyalty, on the other hand, is anchored in trust, empathy, and shared values, qualities that endure beyond transactional rewards.

Rick Delisi, co-author of *The Effortless Experience*, shared a crucial insight with software firm Zendesk: delight doesn't drive loyalty, simplicity does. A study by The Corporate Executive Board (CEB) revealed that loyalty grows when brands consistently deliver on basic promises and effectively address everyday problems. Delisi highlighted that the real loyalty test arises when issues emerge; by resolving these efficiently, businesses can prevent loyal customers from turning disloyal. CEB's research, based on more than 125,000 customers, also showed that customer service interactions are four times more likely to increase disloyalty than loyalty. In response, companies like Mindbody, a software company, are shifting to prioritize problem resolution over simply "wowing" customers.

True loyalty's future lies in a thoughtful integration of technology and humanity. AI should enhance customer understanding and support, not replace genuine connection. When brands use AI to deepen empathy, understand their customers better, and offer proactive support, they build a loyalty that's resilient and authentic.

Ultimately, loyalty is a reflection of shared values and mutual respect, a testament to the strength of a brand-customer relationship. While AI can support this journey, it's the human touch, the assurance that a brand genuinely cares, that transforms loyalty from something fleeting into something lasting.

Moral Intelligence vs. Ethical AI

Moral intelligence enables humans to navigate ethical dilemmas with fairness, compassion, and sensitivity to societal and cultural values. By considering the broader context, humans build trust and loyalty through ethical consistency and thoughtful decision-making that resonates personally with others.

Ethical AI is designed to mitigate bias and ensure fairness by aligning decisions with predefined ethical frameworks. While it excels at systematic evaluation and efficiency, it struggles with the nuances of morality. Unlike humans, Ethical AI cannot adapt to evolving norms or empathize with those affected by its decisions, risking missteps in complex moral scenarios.

Moral Intelligence and Ethical AI
for Customer Loyalty

- **Human-Centric Ethical Governance:** Create hybrid teams of humans and AI to craft and implement ethical standards. Ethical AI can identify biases or inconsistencies efficiently, but human moral intelligence ensures decisions reflect empathy and fairness. This partnership fosters trust by showing customers that businesses prioritize people over algorithms.

- **Evolving Ethical Frameworks:** Design AI systems with adaptable ethical guidelines, updated by human experts, to stay aligned with changing societal norms. Customers value companies that demonstrate an ongoing commitment to fairness and inclusivity.

- **Scenario-Based Ethical Training:** Equip Ethical AI with real-world examples curated by human moral intelligence to handle diverse, context-rich situations. Demonstrating AI's thoughtful application of ethics reassures customers that decisions consider their well-being and individual needs.

Chapter 10

DYNAMIC PROCESS DELIVERY
Practical Intelligence in Customer Operations

Practical intelligence, often thought of as "street smarts," is what happens when experience and instinct collide to solve real-world challenges. It's not about acing exams or memorizing theories; it's about navigating the messy, unpredictable moments where something needs to be done—and fast. In business, practical intelligence matters because operations are rarely as tidy as they appear on a spreadsheet. There's a world of difference between the plan on paper and the reality on the ground, and practical intelligence bridges that gap. It's the ability to make judgment calls that can't be calculated, to act when there's no perfect answer, and to adapt in real time when things inevitably go sideways.

What makes practical intelligence valuable in business operations is its mix of adaptability, social skill, and self-awareness. Take problem-solving, for instance: the magic of practical intelligence lies in knowing that a perfect solution doesn't exist, and that often the best answer is simply the one that gets the job done. It's about seeing patterns, assessing risks quickly, and deciding when to pivot. And in a high-stakes environment, the real trick isn't just finding a solution; it's finding a solution that can work right now. This is especially true in areas like customer service, logistics, and supply chain management, where an unexpected delay or resource shortage requires an immediate, practical response.

Beyond being savvy about processes, practical intelligence also thrives on social acuity. In operational roles, success depends on collaboration, and people aren't predictable variables. Someone with practical intelligence can read a room, manage interpersonal dynamics, and communicate in a way that not only motivates but also gets people moving in the right direction. Practical intelligence doesn't just solve problems; it understands people. Leaders with practical intelligence inspire trust and loyalty, creating a space where teams feel encouraged to speak up, suggest improvements, and work toward shared goals.

In business operations, practical intelligence has a unique relationship with resilience and adaptability. Operations never go entirely as planned, and when a setback hits, practical intelligence becomes a kind of superpower. Those with high levels of it don't panic. Instead, they recognize that obstacles are inevitable and view each one as an opportunity to pivot and innovate. Adaptability in this sense isn't simply a skill; it's a mindset, a readiness to adjust strategies and make tough calls when the unexpected happens. When thinking about industries driven by constant change, like tech or logistics, this flexibility isn't a luxury; it's actually a necessity.

Practical intelligence also drives innovation. It goes beyond maintaining the status quo, seeking out unconventional paths and unique solutions that others might overlook. For companies aiming to stand out, fostering practical intelligence in their teams creates a culture where fresh ideas aren't just tolerated but actually welcomed. Think about a manager who sees inefficiencies in a supply chain. Instead of waiting for a six-month audit, they identify a solution, implement it quickly, and keep refining it. They aren't waiting for an ideal scenario; they're making things better, day by day. This kind of continual improvement and real-time problem-solving can push an entire business forward, creating a culture of action and adaptability that is rare yet powerful.

The exciting thing is that practical intelligence isn't just innate; it's teachable. Companies can cultivate it by prioritizing experiential learning, mentorship, and cross-functional training that helps employees understand the business beyond their specific roles. Learning by doing and having mentors who share real-world wisdom encourages employees to make connections, ask better questions, and, ultimately, make better choices. Practical intelligence doesn't come from memorizing a manual; it comes from exposure to a variety of situations, perspectives, and lessons from those who've been in the trenches.

Practical intelligence has become a competitive edge, especially in today's world where businesses are defined by their ability to adapt. Companies that invest in cultivating this skill in their teams are setting themselves up for long-term success. They're not just responding to challenges; they're also cultivating resilience by embracing flexibility, quick thinking, and a strong dose of ingenuity. It's a type of intelligence that doesn't merely help a business survive; it drives it forward, keeps it evolving, and ensures that it remains relevant in a landscape that is always changing.

The Gap AI Can't Bridge

Albert Einstein said, "*The measure of intelligence is the ability to change.*" That simple sentence describes the profound limitation AI faces in customer operations today. While artificial intelligence is undeniably reshaping business, it's also missing one key ingredient: adaptability. Real, human adaptability. AI can crunch numbers and analyze patterns with lightning speed, but it stumbles when faced with the unexpected twists and subtle shifts that define real-world customer interactions. It's this challenge—the need for fluid, responsive intelligence—that underscores the enduring need of practical human skills in business. According to Deloitte's 2024 Workplace Skills survey, the enthusiasm for AI has overshadowed some critical human capabilities—despite 87% of workers recognizing adaptability, leadership, and communication as key to their career advancement. And yet, in a troubling disconnect, only 52% believe their company truly values these skills.

AI brings efficiency, speed, and data-driven insights, but it falls short when it comes to real-world problem-solving—the kind that relies on intuition, empathy, and the ability to pivot quickly. Workers know this, and they're calling for support in sharpening these human skills, knowing that without them technology-driven efficiency can only take us so far. Deloitte's survey further revealed that 94% of respondents worry that future generations may miss out on developing these critical human skills. The irony is stark: as AI grows more pervasive, the gap between what it can automate and what human intelligence can achieve becomes ever clearer.

Practical intelligence is not just a skill set; it's the competitive edge that AI can't replicate. *Harvard Business Review* data shows that companies that emphasize practical problem solving see a 40% boost in productivity along-side a 30% drop in errors and rework. Why? Because re-al-world problem solving involves not only quick think-ing but also the ability to anticipate, to understand that obstacles aren't static and that operations require ongoing, often intuitive, adjustments. Elon Musk put it succinctly: "Intelligence is not just solving problems but anticipating them." And that's where human intelligence stands apart: knowing when a storm is brewing and heading it off before it disrupts the system. That's the sort of insight AI can't mimic because it's grounded in experience, judgment, and the keen perception of subtle cues that data alone cannot reveal.

The ability to solve problems in real time under re-al-world constraints is where human skills shine brightest. Research from MIT Sloan demonstrates this: companies that prioritize hands-on problem solving outperform their competitors by 45% when navigating disruptions. It's not the flashiest skill, but it's the one that keeps a business alive and thriving when turbulence hits. Human intelligence doesn't just act on data; it acts on nuance. It sees around corners, picks up on patterns that aren't readily visible, and knows when to change course without waiting for perfect information.

In this way, practical intelligence fills in where AI leaves off. It provides the adaptability, foresight, and social intelligence that bring business operations to life. It bridges the gap between what can be predicted and what can't. And where change is the only constant, it's the businesses that cultivate this human element—this irreducible, unscript-ed intelligence—that are setting themselves up for lasting success. Practical intelligence, then, isn't just a set of skills; it's a philosophy that embraces the unpredictable nature of business and turns it into an asset. In an AI-driven world, that philosophy just might be the most essential of all.

Experian's Center of Excellence

Experian's journey to build a Center of Excellence (CoE) for onboarding their teams and partners wasn't just about streamlining processes; it was also about creating a foundation that would empower both customers and teams to thrive amid change. When I began consulting with them through my role at Google, the goal appeared simple enough: craft an onboarding experience that was intuitive, supportive, and scalable. But as we delved deeper, it became clear that Experian wanted more than an efficient process. They sought an ecosystem built to adapt, a system that could meet the challenges of today and evolve to support the innovations of tomorrow.

Working alongside Experian's team, we defined, designed, and developed this CoE with a customer-first mindset at its core. Resources and standardized processes formed the structural backbone, but the heart of the project lay in a dynamic process delivery framework that could flex and adapt. Experian's landscape was changing fast; with new acquisitions came the need to quickly integrate and onboard entire teams. One week, I was consulting with their team on-site in Madrid, the next in London. Each location had its own unique requirements, but our vision remained singular: create a seamless experience that instilled confidence from day one.

The results were striking. In less than three months, Experian's CoE was fully operational—an incredible pace that spoke to the power of our collective focus and ambition. Nine months later, Experian was onboarding teams globally, rolling out a cohesive system that wasn't just user-friendly but was also ready to meet the company's rapidly evolving needs. Every decision in that build was driven by a central question: Will this equip the next team or customer to thrive here, regardless of the changes that lie ahead? As the CoE scaled, the fintech industry took notice. As of 2024, Experian earned recognition as a Global Fintech Leader for five consecutive years—an honor that underscored its commitment to innovation, operational excellence, and customer experience. Industry experts were citing customer experience, advanced analytics, regulatory compliance, and fraud prevention as top fintech opportunities, and Experian was leading in every category. Kathleen Peters, Experian's Chief Innovation Officer, summed it up perfectly: "We're gratified that our success is consistently recognized...and we look forward to continuing to deliver the same standard of data and technology excellence to our customers."

What Experian achieved went beyond just onboarding and process management. With a "global-first" mindset, they were able to introduce products and capabilities across markets faster and with more impact. We didn't just build a cloud-based system; we created an operational ecosystem that made agility, quality, and customer delight central. Experian wasn't simply focused on reducing costs to optimize; they were intent on reallocating resources to fuel ongoing innovation. By making product teams more versatile and engaged, Experian's platform became more than an onboarding system; it became a driver of culture and adaptability.

Since my time at Experian, I have leveraged this case study and the core principles of the CoE at dozens of other companies, each of which has built similar teams to foster delivery and a customer-first mindset. The real impact of Experian's Center of Excellence lies in how it redefined onboarding as an experience of empowerment and connection. Each new team or customer welcomed through this system isn't just brought into the fold; they're given the tools and confidence to succeed. This journey, which began as a consulting project focused on practical solutions, evolved into something bigger: a foundation of adaptability and excellence that positions Experian to meet future challenges head-on. It wasn't just a platform; it was a cultural transformation, a CoE that ensures every customer and team feels not only welcomed but truly prepared for what's next.

The Promises of Applied AI

Applied AI is a marvel of modern technology. It can optimize delivery routes for logistics companies, automate scheduling for service-based businesses, and streamline inventory management for retailers. In the customer-service world, it powers chatbots, automates ticket prioritization, and ensures that the right cases are routed to the right agents at the right time.

Its appeal is undeniable. Applied AI operates 24/7, doesn't require coffee breaks, and scales effortlessly. For businesses managing high volumes of customer interactions or logistical complexities, it's like hiring an army of tireless, precise workers. And in environments where efficiency is paramount (like warehouses or call centers), Applied AI shines. It finds patterns in mountains of data, predicting when demand will spike, which agents are most effective, and where bottlenecks are likely to occur.

But here's the rub: Applied AI isn't human. And customer operations are inherently human. The very nature of customer interactions—filled with emotion, unpredictability, and nuance—means there are moments when an algorithm simply won't cut it. AI, for all its efficiency, operates within predefined parameters. It follows the rules it's given, crunches the data it has, and delivers outcomes based on patterns it recognizes. But what happens when the rules don't apply, the data is incomplete, or the situation deviates from the norm?

Humans with practical intelligence excel in these moments. They can think on their feet, improvise, and adapt to chaotic or unpredictable circumstances. Consider the following scenarios:

1. **The Unexpected Request:** A customer calls with an unusual issue—a combination of circumstances that doesn't fit neatly into the system's predefined categories. The AI flags it as "low priority" because it doesn't understand the nuance. Next, a human agent picks up on the urgency in the customer's tone and escalates the issue.

2. **The Emotional Dimension:** Applied AI can identify frustration in a customer's voice or detect negative sentiment in an email. But it doesn't feel the tension in the room or recognize when a kind word might diffuse a situation. Humans, drawing on empathy and practical intelligence, can de-escalate and connect.

3. **The Unforeseen Crisis:** A major logistical failure occurs—a shipping truck breaks down or an airline faces weather delays. AI might propose rerouting or rescheduling, but humans on the ground make quick, creative decisions to ensure that stranded customers aren't left in the lurch.

These moments aren't exceptions; they're the rule in customer operations. And while Applied AI plays an invaluable role in smoothing the predictable, repeatable aspects of operations, it struggles in the face of ambiguity.

Applied AI in Customer Operations

When it comes to customer operations, companies often leap at AI as though it's the holy grail of solutions. AI will get rid of the redundancies, improve communication, even predict needs before customers know they have them. And it's true that AI can, and does, handle many of these tasks well. But let's get something straight: AI lacks adaptability. It has no street smarts, no instinct, and no intuitive understanding of the unpredictable, ever-shifting nature of human interactions. That's where practical intelligence steps in—not as a competitor but as the essential counterpart.

Applied AI offers unprecedented opportunities for businesses to reimagine customer operations. Yet, for all its strengths, it is not a panacea. While AI can deliver efficiency and accuracy, it cannot replace the uniquely human qualities that build lasting relationships. The real promise of Applied AI lies in its ability to complement human ingenuity, not substitute it.

Imagine a logistics system that not only tracks packages but also predicts potential delivery disruptions before they happen. Applied AI can assess patterns, identify risks, and optimize routes in real time. Yet, when an unexpected variable—a sudden weather event or a misplaced shipment—occurs, the human touch is required to navigate the nuances AI cannot anticipate. AI might offer insights, but humans translate those insights into action that resonates with customers on a personal level.

In customer service, AI-driven tools can analyze historical interactions to predict customer needs. For example, a system might suggest resolutions based on past queries, enabling faster support. But it's the human agent who listens empathetically, reads between the lines, and adapts to the unspoken needs of the person on the other end of the call. Applied AI accelerates the process, but it's the human connection that ensures the experience feels genuine and memorable.

Breaking Down Applied AI vs. Generative AI and Agentic AI in Customer Operations

Applied AI, Generative AI, and Agentic AI are all significant players in the AI landscape, each with unique strengths and purposes. However, when it comes to customer operations, Applied AI stands out as the most reliable and effective option. Here's why—and how it differs from the others:

1. Applied AI: The Efficiency Optimizer

What It Does:

Applied AI specializes in addressing specific, practical challenges, such as optimizing logistics, automating repetitive tasks, or streamlining workflows. It is designed to work within clearly defined parameters, offering precision and speed.

Why It Works for Customer Operations:
- **Focus on Tangible Solutions:** Applied AI excels at handling structured tasks, such as routing calls, managing supply chains, and providing real-time operational insights. It doesn't attempt to simu-

late creativity or adapt beyond its programming, making it a reliable tool for predictable operational needs.

- **Scalability Without Overreach:** Applied AI integrates seamlessly with existing systems, managing high volumes of routine tasks without losing consistency.

Limitations:

- Applied AI lacks the ability to adapt effectively in unforeseen scenarios, as it operates strictly within the parameters of its programming and predefined data.
- It is unable to make decisions that require emotional nuance or a deep understanding of complex contextual factors, which are often essential for building trust and resolving intricate customer issues.

2. Generative AI: The Creative Engine

What It Does:

Generative AI creates new content, such as text, images, or code, based on input prompts. Tools like ChatGPT and DALL-E fall into this category, using algorithms to generate novel outputs.

Why It's Less Suitable for Customer Operations:

- **Focus on Creation Over Efficiency:** Generative AI shines in creative contexts such as marketing campaigns or brainstorming sessions, but it lacks the precision and reliability needed for day-to-day operations. For instance, a generative system might craft an impressive marketing email, but it is not built to optimize the logistics of delivering a product to a customer.

- **Inconsistent Results:** Generative AI occasionally produces outputs that are inaccurate, irrelevant, or contextually off. In a field like customer operations—where precision and predictability are critical—this inconsistency can lead to errors or confusion.

When It's Useful:

Generative AI can play a supporting role in customer-facing operations such as drafting email templates for customer inquiries or suggesting creative responses to specific scenarios. However, it is not the backbone of efficient operations.

3. Agentic AI: The Decision-Maker

What It Does:

Agentic AI mimics decision-making processes, adapting dynamically to changing environments and user-defined goals. It aims to simulate autonomy, making decisions within a framework of parameters.

Why It's Less Suitable for Customer Operations:

- **Over-Engineered for Simplicity:** Agentic AI thrives in highly dynamic, decision-heavy environments but can be overkill for routine customer operations. For example, while it is able to manage the flow of a complex manufacturing process, it may not be the best fit for routing customer calls or resolving common issues.

- **Feels Mechanical:** Despite its adaptive nature, Agentic AI lacks the empathy and intuition needed for trust-building interactions. Customer operations often require personal touches that Agentic AI cannot authentically provide.

Where It Excels:

Agentic AI can be useful for higher-level customer strategy such as modeling the impacts of various policy changes on customer satisfaction. However, it is less practical for handling day-to-day operations where Applied AI's focus on efficiency shines.

Why Applied AI is the Preferred Choice

Applied AI excels in customer operations by delivering predictability and consistency, making it ideal for tasks like managing queries, tracking orders, and automating appointment scheduling. Focused on execution rather than creation, Applied AI ensures accuracy and reliability—essential for maintaining smooth customer interactions. It handles the workload efficiently in high-volume scenarios (such as routing thousands of calls or optimizing delivery routes), proving invaluable for scaling operations.

However, its limitations become evident in chaotic or unpredictable circumstances. Lacking human ingenuity and adaptability, Applied AI struggles to adjust when situations deviate from predefined parameters. By contrast, humans with practical intelligence can quickly improvise, apply intuition, and find creative solutions, which are critical when customer needs fall outside standard processes or unexpected challenges emerge.

The true strength of this technology lies in how it complements human efforts. By automating repetitive and time-consuming tasks, Applied AI allows teams to focus on areas requiring emotional intelligence and creativity (such as addressing complex issues or fostering trust.) Together, this partnership between machine efficiency and human adaptability creates a balanced framework for customer operations that prioritizes both precision and flexibility.

Practical Intelligence vs. Applied AI

Practical intelligence equips humans with the ability to creatively and intuitively solve real-world problems, particularly in unpredictable or dynamic situations. It thrives on adaptability and improvisation, enabling people to navigate challenges with contextual awareness and a deep understanding of underlying issues.

Applied AI excels in structured environments, efficiently managing logistics, automation, and data-heavy operations. Its strengths lie in executing repetitive or predefined tasks with precision. However, Applied AI struggles to adapt when faced with chaos or unforeseen variables—situations where human ingenuity are essential.

Practical Intelligence and Applied AI for Customer Operations

- **Continuous Operational Learning:** Use Applied AI to identify patterns in operational workflows and human decision-making. This creates feedback loops that refine processes while leaving space for human problem-solving in nuanced or first-time scenarios.
- **Real-Time Crisis Management:** Implement Applied AI for rapid data analysis during emergencies such as detecting trends in customer complaints or

operational breakdowns. Human teams can then use these insights to craft agile, context-sensitive responses that maintain customer trust.

- **Innovation-Driven Operations:** Applied AI's ability to streamline repetitive tasks frees human talent to focus on strategic improvements. Teams can concentrate on innovation—designing processes or products that address deeper customer problems—and fostering operational excellence rooted in creativity and context.

Chapter 11

THE UNITY IMPERATIVE
Collaborative Intelligence in Customer Alignment

The aspirational approach to business is inherently customer-centric. That is, a company cannot effectively align its own values with those of its customers if the company doesn't consistently place the customer expectations front and center. McKinsey (2020) stresses the importance of such a customer-centric mindset at all levels of the company:

> From the C-suite to the front lines, employees must feel confident that they have the necessary skills and tools to deliver the best possible experience for customers. Leading companies develop academies that combine digital courses, live workshops, and ongoing nudges to support the development of new capabilities. Each learning journey is tailored to a specific role within the organization. Front-line employees receive practical tips on how to put themselves in their customers' shoes and demonstrate empathy when engaging with

customers. CX managers and innovation teams build skills around redesigning customer experiences and mobilizing cross-functional teams. Executives receive tips on how to support, accelerate, and celebrate customer centricity within the organization, by integrating design thinking into the early stages of the strategic planning process.

The fruits of this customer-focused approach are abundant: According to McKinsey, "[b]rands focusing on customer experience witness 15 to 20% increases in sales conversion rates, 20 to 50% declines in service costs, and 10 to 20% improvement in customer satisfaction."

Attaining Alignment

Over the next three chapters, we'll explore in more detail the ways your business can develop alignment between the values of the business and those of your customers. For now, let me preface that discussion by stressing that a trustworthy understanding of your customers' aspirations must be data driven and fact based. To achieve the requisite level of objectivity, you'll want to turn to analytics, data, and AI. As pointed out in the McKinsey (2020) article,

> Many of the best companies use quantitative research and statistical analysis to ground their decisions in facts about what customers value so they can prioritize the experiences that matter most. The most advanced companies use advanced analytics to run simulations of the expected impact of their potential investments so they can build compelling business cases for them.

The objects of this analysis can include customer surveys, purchase histories, and interaction metrics—all of which can offer invaluable insights into consumer behavior. Advanced analytic tools can be used to parse survey results, social media engagement, and online customer behavior to create a nuanced profile of customer values. Leveraging big data can additionally help you identify broad patterns and trends that may be shaping the identities to which your customers aspire. Further, AI tools can be used to offer predictive analytics for forecasting future consumer behavior, enabling proactive rather than reactive strategies. This overall data-driven approach allows your alignment to be supported by tangible insights rather than mere assumptions, and it lays the groundwork for your brand to take precise steps to both address immediate needs and help your customers fulfill long-term goals.

One other technique that can contribute meaningfully to an understanding of your customer base is competitive benchmarking. Knowing what your competitors are offering can be a window into your customers' values, both in terms of the insights into customer aspirations your competitors may have already gained as well as the market environment that is, in part, contributing to and shaping those values. However, this competitive analysis should go beyond merely features and pricing; it should also look at the aspirational image competitors portray through their branding and marketing. Understanding how your product stacks up against this measurement can help guide your strategy, from product development to marketing messages.

Customer Value Alignment

One morning, I found myself in a conference room at Johnson & Johnson, facing a dozen sharp minds from their data and innovation team. I was asked to guide their team through a half-day workshop exploring alignment, both within the organization and, crucially, with their customers. I planned to facilitate an exercise from my book, *Customer Transformation*, called Customer Value Alignment (CVA).

The exercise is based on a two-sided view of value. On one side, the customer's perspective, and the other, the business's internal teams. I divided the full group into individual teams and handed each participant a card labeled with a random role: CEO, CTO, VP of Innovation, Sr. Director of Sales, Marketing Manager, and Data Scientist. They would answer five alignment questions from the point of view of those roles. Each question prompted them to rate the organization's approach on a Goldilocks-style scale: too much, too little, or just right. An example of a completed form is below.

Value Alignment Scale

Not Enough					Just Right				Too Much	
-5	-4	-3	-2	-1	0	1	2	3	4	5

Example: Major Misalignment

Customer	S.H.	CxO	VP.	Dir.	Mgr.	Empl.	Prtn.
-3	-1	2	3	1	0	0	-3

Example: Strong Alignment

Customer	S.H.	CxO	VP.	Dir.	Mgr.	Empl.	Prtn.
0	0	1	0	-1	-1	0	0

As we began, the room filled with shuffling papers and murmurs of discussion. I could tell some participants were wondering if I was familiar enough with their organization, but I kept my smile and strolled around, listening to each team's debates as they tried to answer these deceptively simple questions. Their confusion was part of the process, part of peeling back the layers.

Halfway through, I saw insights take shape. Teams wrote their alignment numbers in the CVA charts on easels which eventually told a consistent story. Each team shared their feedback, their heads nodding at one another's identical results. Yet, no one quite saw the whole picture.

I explained, "Look at these scores: every team broadly assessed the organization's alignment with customers in the same way—misaligned with your internal teams. You all see the same gaps, but no one has taken action to close them."

Silence fell. Then a senior VP spoke up. "This is powerful." Heads nodded around the room. The insight was clear: despite shared internal perspectives, value alignment must begin with the customer. Internal alignment can be managed and optimized, but true customer-centricity demands more. It requires an organization to adjust its operations to meet customers where they are, not to force the customer to fit into its current processes and structures.

At that moment, the room saw it: customer alignment first, cultural alignment next. This is the foundation of a customer-centered organization. Collaboration becomes an extension of that focus: a collaboration with the customer, for the customer, and ultimately, for the organization as a whole. The unity imperative drives this vision, reminding us that true success stems from aligning individual and collective goals, creating harmony across all stakeholders.

The Consequences of Misalignment vs. Alignment

This workshop experience helped me realize that our aspirations are not isolated from others and can actually influence one another in ways that are either productive or unproductive. The unproductive effects often emerge when values are misaligned, underscoring the importance of alignment in managing aspirations and determining how our values intersect with those of others. From a business perspective, misalignment between a company's values and those of its customers can lead to significant losses.

Educational settings highlight the importance of alignment: A student's expectations and those of an academic institution can significantly impact their success. For example, a student may aim to complete a degree while working full-time. While this may work for some, others might find the institution's expectations for a "quality" education—such as time-intensive coursework—at odds with their personal goals. This creates a misalignment between the student's priorities and the institution's requirements, illustrating how conflicting values can hinder achievement.

By the same token, when personal aspirations do align with organizational goals, there may be a positive impact not only on the individual's achievements but on the organization's capabilities as well. An employee is far more likely to be successful if their career aspirations align with the mission and objectives of the company for which she works, and a company with many such "aligned" employees in their workforce is more likely to grow and succeed over time.

Misaligned Values

How, then, can values be or become misaligned? Perhaps the most obvious misalignment arises from values that are in direct conflict, such as when one partner in a relationship deeply desires raising a family while the other is firmly against having children. However, not all value misalignments are so apparent: tension can emerge when one partner prioritizes support for the other but does not see that support reciprocated. Misalignment may also occur when one individual prioritizes values that are unrealistic or disconnected from everyday life, such as placing excessive importance on external validation or material success. These situations can test the patience of an otherwise supportive partner, especially when such values lead to neglect of key relationships or appear more influenced by societal expectations than authentic personal conviction.

In a business context, value misalignment can similarly create challenges, whether between management and employees or between a company and its customers. For instance, when a company's publicly expressed values conflict with those held by a significant portion of its customer base (as seen in examples like the Bud Light controversy of 2023), the fallout can be swift and damaging. Internally, misalignment between employees' values—such as fair compensation, work-life balance, or professional growth—and management's focus on maximizing profits or efficiency can lead to workplace dissatisfaction, tension, or even walkouts and mass resignations. Addressing such conflicts requires open dialogue, mechanisms for honest feedback, and a willingness to adapt, fostering alignment and mutual understanding to prevent further discord.

Adaptability of Values

Meta's policy changes, announced by Joel Kaplan in January 2025, signal a shift toward aligning the platform more closely with its users' expectations and values. By ending the third-party fact-checking program and introducing a Community Notes model, Meta aims to empower its community to provide additional context for potentially misleading posts. Kaplan highlights the influence and alignment of X's similar program, emphasizing transparency and diverse perspectives to minimize bias. Additionally, Meta plans to ease restrictions on politically sensitive topics while focusing enforcement on illegal and severe violations, acknowledging that its overly complex systems have led to frequent errors and user frustrations.

The move to a more personalized approach to political content is another key change. Meta will allow users greater control over the type and amount of political content they see, reflecting its effort to balance user preferences with its commitment to free expression.

These changes showcase a company attempting to realign with its customers by listening and responding to their concerns. Meta exemplifies how companies can address misalignments and regain trust by simplifying their policies and prioritizing user-driven solutions. Setting aside politics, regulations, and free speech debates, this case study reminds us that, once again, true success lies in aligning with what matters most to your customers.

Sustainable alignment of values requires adaptability, allowing for the evolution of existing values and the emergence of new ones.

Values naturally shift over time, shaped by personal growth, life experiences, and the dynamics of relationships. For example, as a personal relationship matures, partners who initially valued a child-free lifestyle may find their shared values evolving to embrace the idea of starting a family. Similarly, businesses and institutions must adapt to changing values. A company that once prioritized on-site work for all employees may need to embrace remote work as societal expectations shift and technological advancements redefine workplace norms, as seen during the pandemic.

New values can also emerge, presenting opportunities or challenges to alignment if stakeholders fail to recognize their significance. A couple relocating to a new city for professional reasons may discover new values inspired by the relationships, venues, and services unique to their new environment. In a business context, a successful project collaboration might ignite a value-driven focus on innovation, or enthusiastic customer feedback might prompt a brand to expand its product line to meet evolving expectations.

The ability to recognize, embrace, and adapt to these shifts ensures that alignment remains dynamic and resilient, fostering stronger relationships and sustainable growth.

While it's natural for some values to evolve, especially as an organization undergoes changes in leadership or culture, shifting values that directly contradict those of the existing customer base can lead to significant misalignment. Customers often form strong loyalty to brands based on shared values, and if those values are altered, it can result in confusion, mistrust, and even alienation.

Organizations should recognize that it is typically easier—and far more beneficial—to realign internally to the overarching values of their customers than to attempt to shift their values and hope customers will follow. Customers are rarely inclined to adapt their beliefs or priorities to align with a company's new direction. Instead, they are likely to disengage and seek alternatives that better reflect their existing values.

Rather than redefining the company's values in a way that risks alienating its core audience, businesses should fo-

cus on understanding and reinforcing the shared values that
have driven their customer relationships. This approach
fosters trust, loyalty, and long-term success. Realigning internally to meet customer expectations not only preserves
these connections but also strengthens the organization's
ability to adapt without losing its identity or the trust of
its audience.

The Necessity of Purpose

Of course, none of this is possible unless your company has not only a clear understanding of your customers' inspired narratives but also a clear understanding of
its own. The company's purpose provides the basis for a
coherent brand narrative, and much has been written in
recent years about the advantages held by purpose-driven
companies. As Deloitte (2019) summarizes,

> Purpose-driven companies witness higher market
> share gains and grow on average three times faster
> than their competitors, all the while achieving higher
> employee and customer satisfaction. Today's consum-
> ers often identify with a brand's purpose, seeking to
> connect at a deeper level even as the brand reciprocal-
> ly aligns with who [the customers] are and who they
> want to be.

Increasingly, the values that shape a company's purpose include the sort of global values with outcomes that
benefit the whole of society at the broadest level. This is
particularly true with younger consumers, as discussed
by senior analyst and writer at GWI, Duncan Kavanagh
(2019):

Our findings revealed that many consumers today make decisions based on how brands treat their people, how they treat the environment, and how they support the communities in which they operate. When companies align their purpose with doing good, they can build deeper connections with their stakeholders and, in turn, amplify the company's relevance in their stakeholders' lives.

In the Forbes article *Best Brands for Social Impact*, Alan Schwarz (2024) highlights consumers' increasing importance on brands' social responsibility. Forbes partnered with customer insights company HundredX to survey over 185,000 consumers, gathering 4.5 million ratings across over 3,000 brands. Evaluating categories like brand values, sustainability, and community support, the research revealed that strong opinions about social impact rose from 22% in 2023 to 27% in 2024—highlighting increased engagement with purpose-driven brands. As HundredX CEO Rob Pace explains, "People care. There's this debate that this stuff doesn't matter, that it got over-hyped. But we're still seeing the consumer, unaided, more likely to call it out this year [2024] than last," reinforcing the value of social impact on customer alignment.

It's not only social good that drives customer loyalty—value alignment can also resonate on a personal level. For an example showcasing values that are more strictly personal but no less important to customers, marketing strategist Laurent Bouty points to Dove:

> Dove, a personal care brand, understood that their customers aspired to real, authentic beauty rather than the unattainable standards often portrayed in the me-

dia. Their "Real Beauty" campaign resonated deeply with customers worldwide, helping the brand build a strong emotional connection with its audience.

In each of these cases, it was crucial that the experience the company aspired to create for their customers aligned closely with the company's authentically held purpose. No two companies' purpose will be exactly alike, so it is important to find the purpose and correlated values that work for your company, as a 2020 McKinsey article reminds us:

> In alignment with their company missions, Nike seeks to deliver inspirational experiences, Starbucks looks to provide experiences that nurture, and BMW seeks to offer the ultimate driving experience. Costco doesn't try to replicate the experience its customers have at high-end retailers but instead provides a no-frills instore experience that reflects its low-cost promise.

In short, clearly and authentically defining your company's core values provides the self-awareness needed to align those values with those of your customers confidently. Achieving this alignment requires two essential steps. First, ensure that these values are genuinely compatible— for example, a fitness brand that prioritizes wellness and inclusivity will naturally connect with customers seeking empowerment on their health journeys. Second, commit to these values through consistent action, such as implementing sustainable practices or advocating for causes that resonate with your audience. This combination of compatibility and commitment builds trust, demonstrating that your brand genuinely shares and supports your customers' aspirations.

Collaboration and Swarm AI

Collaborative intelligence is the ability of individuals or groups to work together effectively, leveraging diverse skills, perspectives, and emotional intelligence to solve problems, innovate, and build strong relationships. It extends beyond simple coordination or teamwork, emphasizing the nuanced human elements that drive meaningful collaboration. Unlike purely mechanical processes, collaborative intelligence thrives on empathy, creativity, and the ability to adapt dynamically. This involves reading the room, understanding context, mediating conflicts, and finding solutions that balance logic and emotion, making it transformative in both personal and professional settings.

In the workplace, collaborative intelligence fosters innovation by uniting individuals with diverse expertise to solve complex challenges, and the culture of an organization plays a pivotal role in determining the level of collaboration achieved. A culture rooted in psychological safety—where individuals feel valued, respected, and free to share ideas without fear of judgment—empowers teams to work cohesively and creatively. When combined with a focus on inclusion, respect, and diversity, such a culture promotes cross-functional teamwork, effective use of virtual tools, and celebration of varied perspectives. These elements build trust, enhance morale, and drive both innovation and retention. By integrating AI tools that support emotional awareness and empathy, companies can further bolster psychological safety and collaboration, creating an environment where individuals and teams reach their full potential.

Swarm AI takes collaboration into the realm of technology, exemplifying the future of collaborative intelligence by enabling autonomous AI agents to work together on complex tasks. As payment platform PYMNTS (2024) explains, OpenAI's Swarm Framework uses decentralized coordination, allowing agents to optimize logistics in real-time, such as inventory management and demand forecasting. Michael Walker, CMO at the AI agent company SmythOS, highlights this potential, stating that these systems can "optimize everything without waiting for human input." Beyond logistics, Swarm AI enhances customer experiences by providing personalized recommendations through real-time behavioral analysis (PYMNTS, 2024).

Unlike traditional AI, which often relies on centralized decision making, Swarm AI leverages distributed networks to process information and respond dynamically. Each agent operates independently, contributing localized data to the system's overall intelligence while reacting to the environment. This results in a flexible, adaptive, and scalable system capable of tackling challenges ranging from optimizing supply chains to managing autonomous vehicles.

Swarm AI is particularly valuable in environments where rapid decisions are critical. For instance, it can analyze vast datasets to optimize resource allocation or identify patterns in customer behavior. While Swarm AI excels in speed and efficiency, it lacks the emotional and contextual understanding that humans bring to decision making. By integrating Swarm AI with human insights, organizations can harness the strengths of collective intelligence while addressing the need for empathy, creativity, and value alignment.

Psychological Safety in the Mix

Collaboration thrives in an environment of psychological safety. When individuals feel valued, respected, and free to express ideas without fear of judgment or repercussion, they contribute their best thinking. However, psychological safety doesn't happen by chance—it is cultivated through an organization's culture. This is where the challenge of introducing AI into collaborative processes, particularly Swarm AI, becomes apparent. While Swarm AI offers immense potential for enhancing decision-making and problem-solving by aggregating collective intelligence, it does not inherently account for psychological safety.

Swarm AI relies on pooling inputs from individuals in real time to derive optimal outcomes. While this approach is inclusive by design, it does not guarantee that participants feel safe to share their unfiltered opinions or dissent from the majority. Without deliberate attention to psychological safety, participants may succumb to groupthink, hesitate to express controversial views or feel overshadowed by dominant personalities, even in an AI-mediated environment.

To integrate psychological safety into Swarm AI systems, organizations must actively foster a culture that encourages open dialogue and respects diverse perspectives. Techniques such as anonymous participation, transparent feedback loops, and the use of facilitators can help. Additionally, AI systems can be adapted to monitor sentiment, detect patterns of exclusion, and nudge participants toward more inclusive engagement.

Swarm AI has the potential to amplify collaboration, but its success hinges on more than algorithms. It requires a foundation of trust and safety where individuals feel empowered to contribute authentically. By combining the strengths of AI with deliberate human-centered strategies, organizations can create environments where collaboration flourishes—not just through technology but through the psychological well-being of those who power it.

Human Connections

At its core, human connection is about the shared values, emotions, and experiences that bind us together. It's what makes us human—the ability to empathize, collaborate, and build meaningful relationships with others. When businesses embrace CVA, they tap into this fundamental aspect of humanity, aligning their values with those of their audience. This alignment fosters trust and loyalty by creating a bridge between what the company stands for and what its customers believe in, deepening the connection.

Collaboration, whether in personal relationships or professional environments, thrives on human connection. When we collaborate, we bring our unique perspectives, emotions, and values to the table, working together toward shared goals. This act of connection goes beyond mere cooperation—it's an exchange of ideas, trust, and understanding that makes us stronger as individuals and as a group.

Even in nature, we see parallels to human connection. Societies of ants or swarms of bees congregate and cooperate, driven by shared purpose. Humans, too, gather in groups, drawn by shared interests and the desire to connect. Yet, our connection transcends instinct—it's shaped by our values, aspirations, and emotions. Human connection is the foundation of our relationships, our communities, and the businesses we choose to support.

Why Collaboration and Value Alignment Matter

At its core, customer alignment is about fostering relationships built on shared values. It's not just about identifying what customers want but also understanding why they want it and aligning your business's values with theirs to grow together. Collaborative intelligence is essential in this process, as it allows businesses to actively involve customers as co-creators rather than passive consumers. When businesses align their values with those of their customers, they foster trust and loyalty that extends beyond transactional relationships. This alignment not only meets customer needs but also creates meaningful connections rooted in shared ideals. A few standout examples demonstrate how companies achieve this alignment and turn customers into passionate advocates.

LEGO: Creativity and Community

LEGO exemplifies value alignment through its commitment to creativity and community engagement. By inviting customers to submit ideas for new LEGO sets, the company transforms its fans into co-creators. Winning ideas are celebrated, and creators are rewarded, fostering a sense of ownership and inclusion within the LEGO community. This approach aligns with the values of creativity, collaboration, and fun, resonating deeply with LEGO enthusiasts of all ages. Through initiatives like LEGO Ideas, the company builds a vibrant ecosystem where its customers feel valued not just as consumers but as contributors, strengthening long-term loyalty.

Tesla: Sustainability and Advocacy

Tesla's customers are more than just drivers; they are vocal advocates for the company's mission to accelerate the world's transition to sustainable energy. Tesla's values—innovation, sustainability, and boldness—resonate with its audience, many of whom are environmentally conscious and passionate about renewable energy. From electric vehicles to solar panels and battery storage solutions, Tesla consistently delivers on its promise of sustainability. Its customers, aligned with these values, often champion Tesla's mission publicly, becoming part of a larger movement for environmental change. The alignment creates an emotional connection that transcends the products, fostering loyalty and advocacy.

Nike: Empowerment and Social Impact

Nike embraces community engagement by aligning its brand with empowerment and advocacy for social causes. Through initiatives like the *Made to Play* program, which encourages youth participation in sports, and campaigns such as *You Can't Stop Us*, Nike amplifies messages of perseverance, equality, and inclusivity. By partnering with athletes who champion social justice and diversity, Nike resonates with customers who value progress and societal change. This approach not only strengthens its connection to communities but also inspires loyalty by creating a shared purpose. Nike's commitment to empowering individuals and driving impact ensures its place as a brand that leads with purpose.

These examples show how value alignment transforms businesses into movements, creating emotional bonds that inspire customer loyalty and advocacy. When companies actively engage with the values their customers hold dear, they build relationships that endure, driving sustainable growth and mutual success.

Collaborative Intelligence vs. Swarm AI

Collaborative intelligence enables humans to connect deeply, resolve conflicts, and innovate through shared experiences and emotional understanding. It fosters trust and teamwork, aligning individuals and groups around shared values and goals.

Swarm AI uses decentralized systems to mimic group coordination, efficiently solving structured problems and analyzing collective input. However, it lacks the nuanced emotional intelligence and creativity required to align with human values or facilitate meaningful collaboration.

Collaborative Intelligence and Swarm AI for Customer Alignment

- **Shared Value Discovery:** Leverage Swarm AI to gather customer feedback and identify shared values or priorities. Human teams can use these insights to align business strategies with customer expectations, fostering a sense of partnership and shared purpose.

- **Co-Innovation Platforms:** Use Swarm AI to aggregate ideas from customers and stakeholders, identifying the most promising concepts. Humans can facilitate workshops or collaborative sessions to bring these ideas to life, ensuring alignment with customer values and aspirations.

- **Value-Driven Partnerships:** Swarm AI can analyze customer behavior to identify opportunities for deeper alignment with shared values, such as sustainability or community engagement. Human teams can then collaborate with customers to design initiatives or programs that create mutual value and strengthen relationships.

Chapter 12

NAVIGATING THE TERRAIN
Naturalistic Intelligence in Customer Management

Between the ages of 9 and 11, my family spent our summers living and working at a camp called NEOSA, nestled along Leesville Lake in Carrollton, Ohio. My parents ran the camp, and I embraced every wild corner of the place, turning camping into a lifelong passion. To this day, I make an annual pilgrimage to a national park to recapture that sense of adventure. As a staff kid, I had perks like access to the staff lounge with its pinball and pool tables, but I rarely got to spend time with campers my age, especially on overnight excursions. One summer, though, a girls' cabin planned a camping trip into the woods, and I begged my dad to join them. Why did I, a 9-year-old boy, want to camp out with an all-girls group? I couldn't tell you, but I came prepared with my tent, sleeping bag, and an eager sense of adventure.

After heading into the woods and getting my tent set up, I joined everyone for the classic campfire routine: hot dogs, s'mores, and spooky stories that we whispered as the fire crackled in the cool forest air. When it was time for bed, I crawled into my tent and drifted off, lulled by the stillness of the night.

When I woke the next morning, something immediately felt off. There was a soft morning fog layer drifting through the trees, and it was eerily silent—the kind of quiet that creeps into your bones. I peeked outside, expecting the usual morning chatter, but all I saw was chaos. Several tents had collapsed, others were nowhere to be seen, and not a soul was around. My tent was the only one left standing. In my 9-year-old mind, that felt like an accomplishment, but only for a split second before I realized the implications. *Where was everyone?*

My imagination kicked in hard. Had some monstrous creature stormed through the camp, scattering tents and stealing campers away into the night? My heart pounded as I cautiously walked through the woods, making my way to the main path, half-expecting to encounter some nightmare beast at any moment. I knew these woods well—I had been exploring them every summer for three years—but now they felt foreign and threatening.

Eventually, I arrived at a field carved out of the forest as a play area. My heart raced as I scanned the clearing, hoping to see someone, anyone. Instead, I spotted something even more terrifying: seven massive turkey vultures perched like sentinels, watching me with cold, unblinking eyes. A resting group of turkey vultures is called a "committee," but to a 9-year-old, they looked like an army. In my mind, these massive birds, each seemingly 10 feet tall, were the culprits, having carried everyone off in the dead of night.

I froze. Then, a few of them spread their wings and took to the air. Panic hit like a tidal wave. I did the original "Run, Forrest, Run," sprinting all the way back to the dining hall, convinced these creatures were after me. By the time I made it, my imagination was a wild concoction of monsters, rainstorms, and human-snatching vultures.

Later, I learned the actual story: a storm had rolled in overnight, and to avoid getting drenched, the counselors had hurriedly packed everyone back to the cabin. In the rush, they'd forgotten about me, still sound asleep in my tent. I'd slept through the entire thing, blissfully unaware as the storm rattled the camp, collapsed tents, and sent everyone scrambling. But somehow, my little tent had stayed upright, unscathed, sheltering me from the chaos around me.

The Terrain to Customer Management

I've always had a deep love for nature—rooted in childhood camping trips and kept alive today through road trips to national forests and camping adventures with my son. These experiences have heightened my appreciation for what experts call naturalistic intelligence, the ability to observe and connect with nature on a deeper level.

"Naturalistic intelligence is an interest or curiosity about nature, the outdoors, and our planet," explains Courtney Morgan, LPCC, a licensed therapist and founder of Counseling Unconditionally. "People with high levels of naturalistic intelligence are drawn to the outdoors and enjoy spending time in nature."

There's something profoundly grounding about being in the wild, yet it also sparks reflection. The more time I spend immersed in nature, the more I notice striking parallels between the natural world and the business landscape. In customer management, much like in the wilderness, navigating the terrain is unpredictable. Customer behaviors shift without warning, and needs or expectations can evolve in an instant, often without clear signals. It's a dynamic environment that demands both intuition and adaptability.

Naturalistic intelligence stands as one of the lesser-discussed yet profoundly impactful types of intelligence. Coined by psychologist Howard Gardner, naturalistic intelligence refers to an individual's ability to recognize patterns in the natural world. People who possess this intelligence are tuned into nature, plants, animals, and weather patterns, often identifying subtle variations and shifts that others might overlook. Naturalistic intelligence doesn't necessarily mean a person is an expert in biology or environmental science, though it can certainly manifest that way. Instead, it's a keen awareness of one's surroundings, a sensitivity to the interconnections between living things and the environments they inhabit—an ecosystem.

Consider historical events and stories where naturalistic intelligence and adaptability were the difference between success and disaster. The novel *The Perfect Storm*, for instance, depicts a real-life account of a fishing crew that faced relentless, unexpected natural forces. Their journey reveals themes central to customer management: taking stock of the environment, adapting to unforeseen conditions, and, at times, shifting course entirely to avoid catastrophe. Just as the crew navigated a convergence of multiple storms, customer managers often face unpredictable forces in the market, customer behavior shifts, or crises. In these moments, naturalistic intelligence allows us to recognize subtle cues, anticipate needs, and adapt strategically, making decisions that a rigid plan or AI-driven model alone might not capture.

Why does this matter in today's world? Because, in an age when we are increasingly disconnected from the natural world, naturalistic intelligence serves as a grounding force. It fosters a deep understanding of not only our planet but also our company's ecosystems of customers. This intelligence cultivates a sense of stewardship and care, which is critically needed in an era marked by human disconnection and environmental degradation.

As Dr. Gladys Barragan-Jason, a researcher at the Theoretical and Experimental Ecology Station, noted in a 2022 study, "New ways of digitally interacting with nature have certainly emerged or increased in recent decades." Her co-author, Dr. Victor Cazalis, a postdoctoral researcher at iDiv and Leipzig University, elaborated:

> "But several former studies show that these interactions have a lesser effect on our sense of connection with nature than direct interaction. The knowledge about these human-nature interactions is crucial, as they are key in the construction of our relationship with nature and our behaviors. We need to maintain a good connection with nature in order to enable the necessary societal transformations of the 21st century. Only then can humanity live in harmony with nature by 2050."

When we honor and develop our naturalistic intelligence, we become more attuned to our customers' "circle of life," equipping us to make decisions that reflect a long-term vision for our shared future. Amid the prevalence of synthetic content online and virtual connections through social media, this intelligence provides a vital counterbalance. It draws attention to our roots and dependencies in the living world, encouraging the creation of ecosystems in business that mirror nature's care and sustainability.

Environmental AI and Customer Ecosystems

Comparing naturalistic intelligence with Environmental AI reveals striking insights into how we might approach customer-centric leadership. Environmental AI, designed to model and optimize natural ecosystems, mirrors the challenges and opportunities businesses face in managing their own customer ecosystems. By observing the parallels between ecological balance and customer management, leaders can gain valuable perspectives on fostering interconnected and thriving networks.

Environmental AI excels at predicting outcomes and providing actionable insights within ecological systems. Sanksshep Mahendra (2024), CIO at Artificial Intelligence+ explains:

> "AI has been instrumental in advancing technologies around environmental monitoring, which plays a key role in managing ecosystems, air quality, and biodiversity. Traditionally, collecting environmental data required large-scale, resource-heavy operations. AI reduces these limitations through automation, producing faster and more precise insights from the data."

However, like other forms of AI, Environmental AI lacks the innate, organic understanding that naturalistic intelligence embodies. A conservationist perceives the nuances of an ecosystem—the interplay between species, the significance of a single organism in the broader web of life, and the intangible rhythm of the environment. This connection is not rooted in algorithms but in years of observation, intuition, and experience.

Customer ecosystems are living networks of relationships influenced by values, behaviors, and external forces.

A purely data-driven approach, akin to Environmental AI, may identify patterns and predict behaviors but struggles to capture the emotional and cultural nuances that bind customers to a brand. It is the business equivalent of tagging plant species without understanding their role in maintaining ecological balance.

Naturalistic intelligence teaches us that adaptive, responsive engagement is key to fostering resilience, whether in a forest or a customer base. Just as a skilled ecologist reacts to shifting weather patterns or the emergence of new species, a customer-centric leader must remain attuned to the dynamic factors influencing their ecosystem. Environmental AI can assist in monitoring trends, identifying risks, and optimizing operations, but the success of the system relies on a leader's ability to weave these insights into a broader understanding of their customers' needs, aspirations, and values.

In practice, this means creating customer ecosystems that reflect the principles of thriving natural environments:

1. **Interdependence Over Segmentation:** Just as every organism in an ecosystem plays a role, customers, employees, and market forces shape a web of interdependent relationships. Leaders must recognize these connections and design experiences that nurture mutual benefit. A thriving customer ecosystem does not isolate individuals into rigid segments but instead fosters a network that supports and sustains itself.

2. **Adaptability Over Rigidity:** Environmental AI can identify threats or opportunities, but it cannot replicate the intuitive adaptability of a skilled naturalist. Businesses must embrace this adaptability,

responding fluidly to feedback, external market shifts, and emerging customer needs. This means moving beyond static strategies and fostering an organizational culture that evolves in real time.

3. **Holistic Insights Over Data Reliance:** Like a conservationist who understands the unseen symbiosis between species, customer-centric leaders must look beyond surface metrics to grasp the underlying dynamics of their audience. While Environmental AI offers a snapshot of the ecosystem's health, it is the naturalist's intuition that uncovers why certain elements thrive or falter. Businesses must blend data insights with human empathy to truly understand and serve their customers.

4. **Sustainability Over Short-Term Gains:** Ecological balance requires sustaining resources, not exploiting them. Similarly, businesses should prioritize long-term loyalty and trust over short-term profit. Environmental AI can help highlight areas of immediate concern, but it is the leader's role to balance immediate actions with the overarching goal of creating a resilient customer ecosystem.

5. **Purpose Over Process:** Nature's beauty lies in its purpose: every element serves a function, creating harmony. Customer ecosystems should be guided by a similar principle. Leaders must ensure that every interaction—whether powered by Environmental AI or human insight—serves a greater purpose of connection, trust, and shared values.

Thriving in the Ecosystem of Customer Relationships

Environmental AI offers a compelling metaphor for business ecosystems. While it provides tools to understand

and optimize systems, it is the naturalist's awareness—the ability to see the invisible threads of connection and respond to subtle changes—that fosters lasting growth. Businesses must not treat customers as data points but instead as participants in a living network, where adaptability and purpose ensure the system's health.

Viewing customer ecosystems through the lens of Environmental AI transforms how businesses understand and influence customer behaviors. By treating customers as interconnected participants, businesses foster a sense of belonging and shared purpose. This approach encourages trust and loyalty, since customers feel valued beyond transactional interactions. Adaptive, empathetic strategies resonate with customers' evolving needs, cultivating authentic relationships that inspire deeper engagement. Recognizing interdependence within the ecosystem leads to personalized experiences that align with customer values, reinforcing positive behaviors. Over time, this sustainable approach drives advocacy, emotional investment, and a community-centric dynamic, mirroring the resilience of thriving ecosystems.

Ultimately, customer management success lies in blending the analytical power of Environmental AI with the organic, human insights that create genuine connection. Leaders who view their customer ecosystems through this dual lens can build not only thriving businesses but also enduring relationships that mirror the resilience and beauty of the natural world.

Cultures in Nature and Within Organizations

Nature reveals that ecosystems thrive through collaboration and adaptability. Each species within an ecosystem plays a role, from the smallest pollinator to the apex predator, creating a dynamic culture where survival and success are interdependent. Similarly, companies are composed of mini-ecosystems—departments, teams, and roles—that must harmonize to achieve shared goals. When internal cultures mirror the interdependence and balance found in nature, organizations create environments that support innovation, collaboration, and resilience. Just as a healthy ecosystem prioritizes sustainability over short-term gains, thriving company cultures must focus on long-term value creation through shared purpose and mutual support.

Your Culture Isn't Customer-First; It's CEO-First

Customer-centricity, much like an ecosystem's balance, requires intentionality and alignment. Yet, many organizations fall short, operating under a CEO-first culture that prioritizes leadership metrics, profits, and internal efficiencies over genuine customer relationships. To foster a truly customer-centric culture, businesses must adopt principles of naturalistic intelligence, learning from the interconnectedness and adaptability of thriving ecosystems.

Signs of a CEO-First Culture

1. **Leadership KPIs Over Customer KPIs:** A CEO-first culture places company metrics—revenue growth, operational efficiency, or shareholder returns—above customer experience. This approach parallels an ecosystem that prioritizes one species to the detriment of others, ultimately destabilizing the whole. A customer-centric culture, in contrast, recognizes that long-term success depends on aligning business metrics with customer satisfaction, retention, and loyalty.

2. **Top-Down Strategy with Minimal Customer Input:** Ecosystems evolve through feedback loops, with every organism influencing the balance. A CEO-first culture disrupts this natural flow, creating top-down strategies that overlook customer feedback. In a truly customer-first approach, the customer's voice shapes strategy, much like how the smallest organism can impact an entire ecological system. Listening to and acting upon customer needs fosters a healthier and more adaptive business environment.

3. **Misalignment Between Departments and the Customer Experience:** Just as species in nature rely on interconnected roles for survival, organizational departments must work cohesively to deliver a seamless customer journey. A CEO-first culture often results in siloed departments focusing on their individual KPIs, neglecting the larger ecosystem of customer experience. When companies align every function around the customer, they create an environment where collaboration drives success.

Customer-Centric with Naturalistic Intelligence

1. **Foster Interdependence:** Natural ecosystems remind us that interdependence strengthens resilience. Encourage teams to recognize how their roles impact the overall customer experience. From marketing to customer support, align every department's goals with a unified vision of customer success.

2. **Prioritize Customer Feedback:** Just as ecosystems thrive through constant adaptation to envi-

ronmental feedback, businesses must evolve based on customer insights. Build systems that not only gather customer data but also empower teams to act on it, creating an adaptive culture that values responsiveness over rigidity.

3. **Create Shared Value:** In nature, balance ensures survival. Similarly, customer-centric cultures thrive by focusing on shared value. Develop initiatives that benefit both the customer and the business, fostering loyalty through meaningful, reciprocal relationships.

4. **Embrace Sustainability in Customer Relationships:** A thriving ecosystem sustains itself over time. In business, this means prioritizing long-term customer relationships over short-term gains. Focus on cultivating trust, providing consistent value, and nurturing customer loyalty as a long-term investment.

By modeling company culture after natural ecosystems and embracing principles of naturalistic intelligence, organizations can transform into adaptive, collaborative, and customer-centric entities. A culture rooted in shared purpose and authentic connections doesn't just support the customer—it actually empowers the entire organization to thrive.

The Cost of Ignoring the Customer

What happens when an ecosystem prioritizes a single species above the health of the whole? The balance falters, resilience weakens, and the system risks collapse. Businesses operate much the same way. A CEO-first culture that elevates internal metrics over customer needs disrupts the natural flow of trust and loyalty. Customers, like members of an ecosystem, are attuned to imbalances. They quickly sense when an organization's priorities are misaligned with their own and move on to environments that better meet their needs.

Consider the overused statistic: *86% of buyers are willing to pay more for better customer experience.* This data point has become the business equivalent of a tree falling in a neglected forest—acknowledged but not acted upon. Companies recognize its significance but often fail to translate it into meaningful action. The real cost of a CEO-first culture isn't just in lost revenue; it's also in the erosion of a thriving customer ecosystem.

In natural systems, complacency leads to degradation. The same is true for businesses. The challenge isn't knowing the importance of customer experience; it's cultivating the actions, values, and systems that sustain a vibrant customer ecosystem over time.

How Leaders Can Shift to a Customer-First Ecosystem

In thriving ecosystems, balance and interdependence define success. Every organism plays a role in sustaining the whole, and leadership emerges naturally from the needs of the system. To shift from a CEO-first culture to a customer-first ecosystem, leaders must take lessons from nature: observe, adapt, and prioritize the collective health of their environment.

In nature, the canopy provides shelter for life below, but it is deeply connected to the roots and soil that sustain it. Similarly, leaders must leave the high vantage point of the C-suite to engage directly with their customers and frontline employees—the roots of the customer ecosystem.

Former Starbucks CEO Howard Schultz's visits to stores mirror how a forest thrives through constant exchange between its highest branches and its soil. By observing the real-world impact of policies and engaging in unfiltered dialogue, Schultz recalibrated his company's strategies to nurture the ecosystem as a whole.

Every CEO should follow this example. Spend time in the field, listening to customer concerns and observing their journeys. Customer understanding cannot be delegated; it requires direct and empathetic involvement from leadership. This kind of engagement builds authenticity, uncovers hidden challenges, and aligns leaders with the heartbeat of the customer ecosystem.

Align Departments Around the Customer Ecosystem

Species do not operate in isolation within an ecosystem; pollinators, predators, and plants work together to maintain balance. Yet in many organizations, departments function as silos, prioritizing their individual goals over the collective customer experience.

To shift this dynamic, leaders must align all functions around the shared goal of customer success. Marketing, sales, and service should work as cross-functional teams, much like symbiotic species, ensuring that every interaction enhances the customer journey.

For example:
- Marketing KPIs should measure how campaigns support customer needs and foster loyalty.
- Sales KPIs should focus on customer retention and satisfaction over conversion rates alone.
- Customer Service KPIs should evaluate how interactions build trust and lasting relationships, rather than prioritizing speed of resolution.

Even the hiring process should reflect this commitment. Evaluate candidates on their ability to think customer-first, ensuring that every role—regardless of function—contributes to the health of the customer ecosystem.

Celebrate Customer Success

In a thriving ecosystem, the success of one species benefits the whole. Businesses should adopt a similar mindset, celebrating customer wins as part of their own success.

Instead of solely highlighting revenue targets or product launches, shift the focus to moments when customers achieve their goals through your products or services. Share their stories, amplify their successes, and make their achievements the core metric of your organization's success. When customers feel valued and supported, they remain loyal contributors to your ecosystem, strengthening the whole.

The transition from a CEO-first to a customer-first culture requires leaders to think like ecologists: nurturing balance, valuing interdependence, and fostering adaptability. By aligning every function within the organization around the customer and celebrating their success, companies create ecosystems that thrive—not just for a season but for generations to come.

In nature, ecosystems flourish when they prioritize connection and sustainability. Businesses that emulate this approach through naturalistic intelligence can transform their customer relationships from transactional to transformative, ensuring resilience, growth, and enduring loyalty.

Software developer HubSpot famously tracks "Customer Love," a metric that captures how much customers appreciate and advocate for their platform. Instead of focusing solely on internal wins, they measure how well they're helping their customers achieve business success.

Making customer achievements a central part of the company's celebrations and storytelling reinforces a customer-first culture and deepens customers' emotional connection with your brand.

Case Study: The Fall of Disney's Customer Focus

The Walt Disney Company once epitomized what it meant to be a customer-first company. Under the visionary leadership of Walt Disney, the company was known for delivering magical experiences, putting customer satisfaction above all else. Walt clearly understood that delighting guests was the key to long-term success. His philosophy was simple: "Do what you do so well that they will want to see it again and bring their friends."

Over time, however, Disney's customer-centric ethos has shifted. Today, many argue the company has prioritized profit over the guest experience, with leadership focusing more on shareholder returns than customer satisfaction. A glaring example of this shift is the consistent and steep increase in ticket prices at Disneyland.

In 2022, single day ticket prices for the highest tier jumped to $179, up from $164 in 2021. By 2023, that price had risen again to $194, an 8.38% increase. The most recent hike in 2024 brought the top-tier ticket price to $206, a 6.2% increase compared to the previous year (MacDonald, 2024). For those looking to enjoy Disney's magic year-round, the lowest-priced annual pass option saw a staggering 20% increase. These price hikes have made the once-accessible Disney experience feel increasingly out of reach for average families.

While Disney's revenue has undoubtedly increased, these decisions have come at the expense of customer loyalty and satisfaction. Fans who grew up with the brand are beginning to feel alienated, perceiving the company as more concerned with maximizing profits than maintaining the magic that once made it special. Frustration is rising as many loyal guests express that Disney's focus has shifted from creating a magical, inclusive experience to simply cashing in on its brand.

I often imagine a two-month competition between myself and Disney's CEO, Bob Iger. In this scenario, I would be given free rein to make any changes I saw fit, and at the end of those two months, we'd compare customer satisfaction scores and revenue. I'd reduce prices, reduce capacity limits in the park, and even close the park on Wednesdays, providing time for everyone to recuperate.

I have no doubts that refocusing on customer delight rather than profits would result in higher satisfaction scores and, ultimately, more significant long-term revenue. By putting the customer first and restoring the magic slowly drained by a CEO-first mentality, I'm confident that Disney could reclaim its once-untouchable status as a brand people love, not just tolerate.

The Bottom Line

Transforming from a CEO-first culture to a customer-first one isn't easy. It requires a fundamental shift in organizations' operations from top-down leadership to cross-functional collaboration. But the payoff is immense: increased loyalty, stronger relationships, and a competitive edge in a crowded marketplace.

It starts with a simple yet powerful mindset change: the customer isn't just another stakeholder in your business; they are your business. Everything else is secondary. If leaders can embrace this truth, they'll create an organization where both the customer and the company thrive together.

Naturalistic Intelligence vs. Environmental AI

Naturalistic intelligence empowers humans to recognize patterns and relationships within ecosystems, adapt to changes, and navigate complex terrains. In customer management, this translates to understanding the interdependencies within customer ecosystems, adapting strategies to their evolving needs and fostering long-term relationships.

Environmental AI processes vast amounts of data to identify trends and make predictions. Applied to customer management, it can analyze behavioral patterns, forecast needs, and suggest strategies. However, AI lacks the intuition to detect subtle shifts or context-specific nuances, which can lead to oversimplified conclusions.

Naturalistic Intelligence and Environmental AI for Customer Management

- **Map Customer Ecosystems:** Use naturalistic intelligence to understand interconnected behaviors and Environmental AI to model dynamic relationships, predicting changes across internal and external ecosystems.
- **Foster Adaptability:** Emulate nature's resilience and employ Environmental AI to adapt customer strategies, offerings, and processes in real-time to meet evolving needs.
- **Build Holistic Insights:** Combine diverse data sources and Environmental AI to understand customer behaviors, considering economic factors for tailored, customer-centric strategies.

Chapter 13

THE INTELLIGENCE PARADOX
Existential Intelligence in Customer Growth

Existential intelligence (ExI) is a concept proposed by psychologist Howard Gardner as a potential ninth intelligence related to the deep and profound contemplation of life's essential questions. Gardner states, "Existential intelligence is the cognitive capacity to raise and ponder big questions, queries about love, about evil, about life and death, indeed, about the nature and quality of existence." Individuals with strong ExI often seek to understand complex and abstract ideas that others may overlook or consider intimidating.

Unlike logical-mathematical intelligence, which centers on concrete problem-solving, existential intelligence leans toward introspection and philosophical exploration as well as "the big picture." Those with heightened ExI might question the meaning behind daily activities, societal norms, or natural phenomena. They find value in examining moral beliefs, spiritual concepts, and philosophical frameworks. Thinkers, theologians, and philosophers such as Socrates, the Buddha, and Jean-Paul Sartre exhibit ExI by questioning established norms and examining human existence from multiple perspectives.

In 2024, Gardner reflected on the increased interest in ExI. He explains that while he has not formally recognized the concept as an independent intelligence due to the lack of clear neurological evidence, recent global crises (like the COVID-19 pandemic) have intensified public curiosity about such deep reflections. Gardner connects this resurgence of existential questioning to societal disruptions, personal reflection time, and the timeless philosophical search for meaning—particularly in the face of life's inherent uncertainties. He notes his own renewed engagement with existential themes through literature, philosophy, and Camus' novel *The Plague*, which underlines life's inherent meaninglessness and the human need to create personal significance, often through compassion and decency toward others.

In practical terms, ExI enables individuals to engage in reflective and meaningful conversations, connect with others on a profound level, and offer perspectives that challenge the status quo. This intelligence can also manifest in art, literature, and science, where creators explore themes that resonate with the human experience on a universal scale.

Although not as easily measurable as other intelligences, ExI holds significant value today. As people seek purpose, mindfulness, and deeper connections with others, existential intelligence offers tools to foster self-awareness, empathy, and resilience. Through the lens of ExI, individuals explore not just how to live, but also how to live meaningfully, navigating the complexities of life with purpose and a greater sense of interconnectedness with the universe.

Existential intelligence isn't merely an abstract trait; it's deeply personal, shaping how we interpret our own stories and see the world around us. For many, the journey of exploring existential themes often begins in childhood, sparked by seemingly small yet powerful moments. For me, these moments subtly ignite a sense of curiosity about life, purpose, and growth. In the simplest routines and the thrill of facing challenges, I often search for ways to reflect on resilience, perseverance, and the idea of growth through effort and experience.

The Donkey Kong Machine

Growing up in West Hempstead, New York, my mornings started early with delivering The Daily News (newspaper) on my BMX bike. I'd glide quietly through the streets, working my way past familiar houses before dawn fully arrived. I was once chased by a dog who seemed about the size of a bear in my memory) just for getting too close to his yard. But what I looked forward to after finishing my route was Sunrise Bagels. I'd swing by the shop, savoring the smell of freshly baked bagels before grabbing a hot, crisp, golden one right when they came out from the oven.

If I'm being honest, the bagels weren't the only draw. Tucked into a corner of the shop was a gleaming Donkey Kong machine, its bright screen glowing like a beacon for kids like me. I spent countless quarters on that game, trying to master the jumps, ladders, and rolling barrels. With each level, the challenges became harder, but that only increased my determination. Some days I'd make progress; other days, I'd barely make it through the first level. But I kept coming back, quarter after quarter, determined to save Pauline.

A year later, Sunrise Bagels replaced Donkey Kong with Donkey Kong Jr., and later with a rotating series of new games. It was an unexpected progression, as though the shop was growing alongside me, presenting new challenges just as I outgrew the old ones. It became a ritual: bagel and quarters. I'd take on each new game, seeing how far I could get before I had to head home and get ready for school. Every attempt presented a new learning curve, a new chance to improve, and I thrived on those challenges—much like the thrill of delivering papers in record time.

During my four-year delivery career, the bagels stayed consistently delicious, but the games kept changing and evolving. The shop didn't just stay the same; it adapted, offering new experiences while keeping the essentials—the bagels, the warmth, the familiar faces—intact. That experience showed me that true growth isn't just about sticking with what works; it's also about making small but meaningful changes that keep people engaged and coming back. Just like the changing games on the Donkey Kong machine, growth for businesses and customers alike means offering something familiar yet new, creating spaces where people can continuously learn, enjoy, and reach for that next level.

Exploring the Push for Human Optimization

In her 2019 article for *Vox*, Sigal Samuel delves into the world of biohacking—a practice that merges science, technology, and self-experimentation in a quest to push the limits of human biology. Biohacking captures a spectrum of activities aimed at optimizing physical and cognitive performance. Some pursue it as a means to enhance health and longevity, while others see it as a radical path toward transhumanism.

Transhumanism (such as bionic limbs and augmented human hearing)was popularized by English biologist and philosopher Julian Huxley in his 1957 essay of the same name, and has been a topic of philosophical debate in recent decades. Britannica (2024) defines transhumanism as:

> A philosophical and scientific movement that advocates the use of current and emerging technologies—such as genetic engineering, cryonics, artificial intelligence (AI), and nanotechnology—to augment human capabilities and improve the human condition. Transhumanists envision a future in which the responsible application of such technologies enables humans to slow, reverse, or eliminate the aging process, to achieve corresponding increases in human life spans, and to enhance human cognitive and sensory capacities. The movement proposes that humans with augmented capabilities will evolve into an enhanced species that transcends humanity—the "posthuman."

At its core, biohacking reflects a universal desire: to feel better and achieve more. Figures like Bulletproof founder Dave Asprey epitomize the movement's ethos, blending supplements, meditation, and futuristic interventions to not just maintain health but also strive for "peak performance." Asprey's approach, as he puts it, involves "changing the environment around you and inside you so that you have full control over your own biology." Similarly, biohackers like Josiah Zayner venture into uncharted territory, injecting CRISPR (Clustered Regularly Interspaced Short Palindromic Repeats) DNA to experiment with genetic enhancements.

While some hacks draw from age-old practices like fasting or meditation, others involve cutting-edge technology. For example, "grinders" implant chips to interact seamlessly with their environments. Critics, however, question the safety, legality, and ethics of these self-driven experiments. The movement's anti-establishment streak often collides with regulatory concerns, as biohackers challenge traditional medicine's pace and processes.

Biohacking, Samuel notes, is more than an individual pursuit; it's a cultural phenomenon reflecting society's evolving relationship with technology and human potential. The movement invites us to question where the line between human and machine should be drawn—and what it truly means to live better, longer, and more fully.

Existential Intelligence and the Fight for Human Agency

The existential threat posed by artificial intelligence lies not in the machines themselves, but in the narrative surrounding them—a narrative that diminishes human agency. As philosopher Shannon Vallor articulates, "The risk I talk about is existential in the philosophical sense of really striking at the core of human beings and our ability to give meaning to our existence." This existential threat is not about AI becoming conscious or malicious; it's about the growing rhetoric that undermines the uniqueness of human cognition and freedom.

Vallor highlights a disturbing trend: the suggestion that humans are no different from AI, merely "predictive text machines." This narrative, she explains, acts as a form of gaslighting, eroding our confidence in human agency. "That's what will enable humans to feel like we can just take our hands off the wheel and let AI drive," she warns. This surrender represents a profound loss of the capacity to shape our futures and make meaningful moral, political, and artistic choices.

Drawing on José Ortega y Gasset's notion of "autofabrication," Vallor emphasizes the uniquely human responsibility to continually create and re-create meaning. "Autofabrication," she explains, "literally just means self-making. This is the fundamental human condition: to make ourselves over and over again." This process requires courage and strength, as it's easier to let someone—or something—else define the script for us. AI, while a powerful tool, becomes dangerous when its narrative persuades us to relinquish this responsibility.

Vallor insists that we must reject the notion that AI's logic is final and unchangeable. "To surrender our human freedom… would mean giving up the possibility of artistic growth, of political growth, of moral growth—and I don't think we should do that." Instead, we must reclaim our agency, seeing AI as a tool to enhance human potential, not replace it.

Human Agency, AI, and Its Role in Customer Growth

Customer growth is not merely a metric of increasing numbers or expanding market share; it's about building relationships that resonate deeply, align with customer values, and adapt to their evolving needs. As Shannon Vallor's insights on human agency highlight, businesses must avoid becoming trapped in rigid systems or scripts, whether those are dictated by AI or outdated practices. Instead, they must embrace the dynamic interplay of consistency and thoughtful innovation to cultivate growth that feels personal, meaningful, and enduring.

True growth strikes a balance: it offers customers reliability while inviting them to explore new and enriching experiences. Consider the example of a beloved café that regularly refreshes its menu or introduces unique events. Customers return not just for the coffee but also for the sense of connection, familiarity, and intrigue the café fosters. Businesses that excel at customer growth understand this principle, offering "next levels" through new products, enhanced services, or deeper ways to connect. This approach makes customers feel like participants in a journey, not just recipients of a transaction.

AI plays a powerful supporting role in this equation. Predictive tools can forecast trends, analyze behaviors, and identify opportunities for innovation. For example, AI might reveal that customers are seeking more sustainable options or craving personalized experiences. However, human intuition is what transforms these insights into actions that resonate emotionally. A machine might identify a demand for eco-friendly practices, but it takes human imagination to weave sustainability into a brand's story in a way that inspires loyalty.

The art of pacing change is critical. Customers appreciate evolution when it feels aligned with their relationship to the brand, not imposed on them. AI can guide the timing of changes by analyzing data, but human agency ensures those changes align with customer psychology and brand identity. This delicate balance fosters loyalty and trust, allowing customers to perceive growth as a partnership rather than a disruption.

Vallor's concept of "autofabrication" is particularly relevant here. Just as individuals continually evolve in order to adapt and find meaning, businesses must also engage in this ongoing process of reinvention. When businesses align their purpose with their customers' aspirations, they create something rare and enduring: a shared sense of identity and mission. This alignment fosters relationships that stand the test of time, grounded in mutual respect and shared values.

Customer growth powered by existential intelligence transcends one-off transactions or quick wins. It's about crafting a dynamic and evolving relationship where customers feel seen, valued, and empowered. When a business operates with purpose and integrates both AI-driven insights and human creativity, it sets the stage for sustained growth that benefits not only the bottom line but also the broader human connection at the heart of commerce.

What is Philosophical AI, and Why Does It Matter?

Philosophical AI represents a distinct branch of artificial intelligence aimed at tackling complex, abstract problems. Bringsjord & Govindarajulu (2024) articulate:

Philosophical AI is AI, not philosophy, but it's AI rooted in and flowing from philosophy. For example, one could engage, using the tools and techniques of philosophy, a paradox, work out a proposed solution, and then proceed to a step that is surely optional for philosophers: expressing the solution in terms that can be translated into a computer program that, when executed, allows an artificial agent to surmount concrete instances of the original paradox. Before we ostensively characterize Philosophical AI of this sort courtesy of a particular research program, let us consider first the view that AI is in fact simply philosophy, or a part thereof.

This definition underscores how philosophical principles can be operationalized within AI systems to address ethical dilemmas, societal impacts, and long-term consequences. Philosophical AI doesn't seek to replace human inquiry but to enhance it, offering tools that help navigate abstract challenges in decision-making and strategy.

Philosophical AI excels in structured reasoning, identifying inconsistencies in policy, exploring the ethical implications of technological innovations, and modeling scenarios in morally ambiguous situations. These capabilities make it particularly valuable in governance, bioethics, and environmental decision-making. However, its limitations are evident: it lacks the introspection, emotional nuance, and deeply personal engagement that define human philosophical inquiry.

As Max Noichl (2024) notes:

> The first idea that we might have when thinking about how artificial intelligence might serve to automate philosophy is that the AI system is going to philosophize for—instead —of us. And indeed the currently best publicly available systems seem to show some basic promise. They are able to recapitulate classic philosophical arguments and thought experiments with reasonable, although somewhat spotty, quality, and when vaguely prompted to opine on topics of philosophical impact they are also able to identify classical lines of argument. But these abilities are very much in line with an understanding of Large Language Models (LLMs) that sees them largely as sophisticated mechanisms for the reproduction and adaptation of already present textual material, which would of course be in stark contrast to the capabilities that are arguably necessary for the production of truly novel and logically coherent philosophy, namely strong abstract reasoning capabilities.

Philosophical AI's true value lies in its ability to integrate and structure vast amounts of philosophical work, fostering cohesion and advancing collaborative efforts. A recent pilot study exemplified this by analyzing thousands of philosophical texts to extract, categorize, and map key arguments and positions. Such systems can identify thematic clusters, consensus points, and areas of contention, enabling faster research synthesis and collaboration.

Imagine an AI capable of parsing thousands of ethical AI papers to create a global map of the field, revealing alignments, conflicts, and gaps in the discourse. This level of analysis accelerates understanding and provides actionable insights, bridging the gap between theoretical inquiry and practical application.

While incapable of original philosophical thought, Philosophical AI complements human insight by organizing knowledge and enhancing decision-making frameworks. It empowers philosophers and businesses alike to focus on generating new ideas rather than revisiting existing debates, making it an indispensable tool for navigating today's complex challenges.

The synergy between Philosophical AI and human intelligence is particularly transformative for business. By addressing profound questions about ethics, societal impact, and long-term strategies, this collaboration ensures decisions resonate with purpose and values. For example, in the realm of customer growth, Philosophical AI can identify emerging trends, anticipate societal shifts, and evaluate the ethical ramifications of scaling strategies. It might pinpoint increasing consumer demand for sustainable practices or reveal reputational risks tied to unsustainable materials.

Yet, this capability is incomplete without human interpretation. Philosophical AI provides the scaffolding, but humans bring the creativity, empathy, and cultural sensitivity needed to craft strategies that truly resonate. Businesses that integrate this partnership can align their decisions with not just data but also purpose, building trust and loyalty.

This collaborative interplay enables businesses to navigate a complex landscape with foresight and integrity. By pairing AI's logical outputs with human values, businesses foster meaningful relationships and sustainable growth, ensuring that progress aligns with the aspirations of both stakeholders and society.

Bread, Bagels, and Building Connections

Jean Valjean stole a loaf of bread to save a child. It's a moment in *Les Misérables* that resonates universally—not because it's a story about theft, but because it forces us to confront a paradox: whose life is more valuable? The bread becomes more than bread; it becomes a symbol of survival, morality, and the complexity of human choices.

Paradoxes like this are more than literary devices; they're lenses through which we see the messy, layered truths of life. And, surprisingly, they also show up in business. Philosophical AI, for example, can analyze paradoxes, simulate reasoning, and provide logically sound answers. But it does so without truly understanding the deeper implications. It can mimic intelligence but not introspection, simulate insight but not experience.

This brings us to the paradox businesses face today.

Our AI Relationship

At first glance, AI seems like a silver bullet for customer growth. It predicts preferences, analyzes behavior and identifies trends with remarkable efficiency. It's the Jean Valjean of business tools, grabbing the loaf of bread—insights and efficiency—to feed a company's hunger for growth. Klarna, the Swedish buy-now-pay-later (BNPL) company, has recently made significant strides in leveraging AI to enhance operational efficiency and financial performance.

Klarna's CEO, Sebastian Siemiatkowski, made bold claims in 2024, stating, "AI can already do all of the jobs that we as humans can do... we stopped hiring about a year ago." On the surface, it's a startling, almost dystopian declaration. The numbers—17% more business volume with 1,000 fewer employees—paint a picture of efficiency at the expense of humanity (Saltmarch, 2024). However, despite these efficiencies, Klarna seeks human talent in specific areas. As of January 2025, the company had 40 engineering roles posted, indicating a strategic approach to human resource allocation. Furthermore, the natural decrease in jobs and increase in profits can be attributed to broader global economic conditions post-COVID during 2023 and 2024.

This rhetoric isn't new. AI-driven hype often leans into fear and urgency: "The window is closing," "Entire industries are vanishing," and "AI can already do all of the jobs." While the numbers provide a veneer of inevitability, they often lack nuance. The question remains: Is the future truly one where AI dominates at the expense of human roles, or is there a middle ground?

Here lies the intelligence paradox: Where do the customers come from if AI replaces employees en masse?

Who will buy the products or services that AI-enabled companies produce if people aren't earning wages? Economic growth depends on customers—customers who need purchasing power.

Henry Ford understood this over a century ago. He paid his employees well, not out of altruism but to ensure they could afford the cars they built. If we pursue a future where humans are sidelined in favor of machines, we risk severing the very economic loop that drives growth. Companies need customers just as much as they need efficiency.

Hyperbole or Truth

Klarna's claim that AI can replace "all jobs" contradicts its simultaneous recruitment of new employees. It reveals a more profound truth: AI may augment or even displace specific tasks, but it does not eliminate the need for human creativity, oversight, and adaptability. In the short term, automation can reduce costs and increase productivity. However, in the long term, a system without humans—both as employees and consumers—cannot sustain itself.

AI's transformative potential is undeniable. McKinsey predicted that by 2034, 800 million jobs will be replaced by AI and automation, with 70% of office tasks automated. Yet, the *World Economic Forum* stated in 2024 that 95% of today's workforce is unprepared for this shift.

But here's the contradiction. IDC (2024) suggests that AI will contribute $19.9 trillion to the global economy by 2030. How do these numbers align? How can there be a loss of income on one side and improved economic conditions on the other? The disconnect between these figures underscores the complexity of AI's role in reshaping the economy.

The Binary Mantra

A suggestion I often see on LinkedIn goes like this: "*By the end of 2025, there will be two types of companies: 1. Those that mastered AI, and 2. Those that don't exist anymore.*"

This oversimplified mantra, frequently echoed by AI proponents, dismisses the deeper complexities of business and the existential intelligence required for sustainable success. It frames the future in binary terms, ignoring the critical paradox: while mastering AI might streamline operations, it risks alienating or displacing the humans businesses rely on for growth. True success lies in technological dominance and balancing innovation with the values, needs, and well-being of customers and employees.

Elon Musk underscored this in April 2023 when he stated, "AI is the biggest risk to the future of civilization." His warning wasn't about AI but how humanity might misuse it—driven by arrogance, ignorance, or shortsightedness. This is where existential intelligence becomes essential. The AI conversation cannot be reduced to an all-or-nothing debate. If it is, the results may very well be nothing.

Machines may process data and perform tasks with unprecedented efficiency, but they cannot empathize, dream, or connect meaningfully. AI can transform our world, but the ultimate decisions about integrating it into society rest with us. Balancing progress with humanity is not just a challenge—it's the responsibility of every leader to navigate this rapidly evolving landscape. AI shapes our future, but human imagination, compassion, and courage define its purpose.

Applying Existential Intelligence to Customer Growth

So, how do businesses escape the paradox?

Start with purpose. What does your brand stand for? How does it align with your customers' values? A brand isn't just a logo; it's a promise. Customers want to be part of something bigger than a product. They want to belong to a mission, a story, a shared identity. Maybe they should be a part of the AI conversation as well.

Examine the customer journey. Are there moments when your brand can offer something meaningful? Not just in the form of a discount, but an actual connection. Not just a product, but a purpose. Think of your customers not as end-users but as participants in a larger narrative.

Use AI as a tool, not a driver. Let it support your understanding of what customers do, but let existential intelligence guide why they do it. AI can map the terrain, but humans must chart the course, drawing on intuition, empathy, and creativity to navigate the emotional landscape of customer relationships.

Customer growth is more than increasing numbers or market share; it's about cultivating relationships that adapt to evolving needs and resonate with purpose. Like a beloved café that refreshes its menu, businesses must balance consistency with thoughtful innovation. This blend creates reliability while inviting customers to explore, engage, and feel valued.

Take Nike's "Just Do It." It's not just a slogan; it's a call to action. It invites customers to believe in themselves, to push boundaries, to overcome challenges. People don't just buy Nike products—they buy into Nike's purpose. That's existential intelligence in action.

Or consider Patagonia, a brand that has embedded environmental stewardship into its DNA. It doesn't just sell jackets; it sells a mission to save the planet. Customers align with Patagonia because the brand aligns with their values.

These brands succeed because they transcend transactions. They focus on the why.

The Secret to Sustainable Growth

Existential intelligence reminds us of the big picture. It's the antidote to the transactional, impersonal interactions that data-driven models often create. Customers don't want to be an equation solved by algorithms; they want to be understood as people.

And people, after all, aren't just looking for products—they're actually looking for purpose. They want brands to reflect their values, to amplify their aspirations, to meet them not just where they are but where they want to be.

The paradox isn't a problem to solve; it's a truth to embrace. AI may bring efficiency, but human intelligence brings meaning. Together, they create growth that isn't just measurable but meaningful—a partnership that builds loyalty, trust, and lasting impact.

In the end, the question isn't how much we know; it's how well we understand. It's how deeply we connect. It's how bravely we ask the big questions and answer them not with data alone, but with purpose, humanity, and heart. That's the real intelligence. That's the real growth.

The Meaning of Growth

Employees expect their jobs to provide a sense of purpose. That expectation isn't a nice-to-have; it's a must-have. A recent McKinsey & Company report highlights the stakes: "Employees expect their jobs to bring a significant sense of purpose to their lives. Employers need to help meet this need, or be prepared to lose talent to companies that will." This isn't about free snacks or remote work flexibility; it's about a fundamental shift in how we think about work, purpose, and the role of leadership in shaping a meaningful workplace.

As it turns out, 70% of employees say their sense of purpose is defined by their job. That means the impact of a role goes far beyond just compensation or benefits. People want their work to matter, to them and to the world around them. According to *Harvard Business Review*, ninety percent of global employees and executives feel that work should bring a sense of purpose. Yet there's a gap. A significant one.

In the same survey, 35% of employees reported a disconnect between their organization's stated purpose and its day-to-day actions. For many, it's a frustrating mismatch between what the company says it stands for and what it actually does. Think about that: more than a third of employees feel like the brand promises they represent are empty words, a façade without substance. That's a huge problem for leaders, because the data couldn't be clearer: When a company's purpose aligns with its employees on a personal level, those employees are 77% more likely to be engaged at work and 87% more likely to want to stay. Translation: Purpose-driven alignment isn't just good for morale; it's good for retention, engagement, and ultimately, growth.

Purpose matters not just for employees but also for customers. Customer growth rate (the speed at which you acquire new customers) is a key indicator of your company's financial health and market demand. A positive growth rate signals that you're tapping into a real demand, and it's a metric that investors, executives, and marketers all track closely. But here's the thing: customer growth isn't just about sales or flashy advertising campaigns. It's about seeing the big picture, connecting the dots between what you offer and what your customers are actually seeking. It's about aligning your company's values with the values of your customers. And just like with employees, this alignment of purpose and values drives loyalty, engagement, and sustainable growth.

Yet AI can't do this. It's brilliant at the science of customer growth, but it struggles with the art of it. It can't answer the "meaning of life"-style questions that lead to deeper understanding and authentic connection. In *The Hitchhiker's Guide to the Galaxy*, Douglas Adams famously posited that the ultimate answer to life, the universe, and everything is "42." But what about the question? AI operates like Adams's supercomputer, Deep Thought— capable of calculating the ultimate answer but incapable of defining the question itself. Sure, AI can predict that a certain demographic is likely to buy a product. But it can't tell you why that product resonates with them, why it makes them feel seen, valued, or even inspired.

The art of customer growth lies in uncovering those deeper questions—the ones that define purpose, meaning, and emotional connection. And that's something AI, no matter how advanced, simply cannot compute.

If customer growth is about acquiring new customers and expanding your market share, then it's also about understanding what draws people to you and keeps them there. A growing customer base is a signal that there's something meaningful in what you're doing. It says, "This matters." But that's where a lot of brands miss the point. They're so focused on the next sale that they forget to ask the big questions: *Why do our customers care? How does our work make their lives better?* These are existential questions, the kind AI just can't tackle, no matter how advanced it gets.

The same existential intelligence that fuels personal growth—questions of purpose, value, and connection—also drives meaningful customer growth. A purpose that resonates isn't just a marketing slogan; it's a magnetic force that attracts people who believe what you believe. And when your employees believe in it too, they're far more likely to go the extra mile to bring that vision to life for customers. That's the magic of alignment. It's the secret to sustainable growth where people are constantly searching for meaning, not just products.

So what does this mean for leaders? It means that, as a leader, you have to do more than manage operations and hit KPIs. You have to help people find their purpose—and live it. Because when your employees are aligned with the purpose of your organization, they're not just working; they're engaged, invested, and driven. And when your purpose aligns with your customers' values, they're not just buying; they're joining you in something bigger.

The companies that stand out are the ones that offer more than just transactions. They offer connections. They offer purpose. They offer meaning. And that's something no algorithm can replace.

People, employees and customers alike want to feel that their work and choices matter. They want to see that the companies they align with stand for something real, something bigger than a sale or a quarterly report. If you can't offer that, if you can't help your people find and live their purpose, you'll lose them to someone who will.

Growth is about alignment. It's about finding the overlap between your values and the values of those you serve, whether they're employees or customers. It's about making sure that what you say matches what you do, that your purpose isn't just a slogan but a lived reality. As people increasingly seek meaning, this approach may be the most valuable asset a business can offer.

Existential Intelligence vs. Philosophical AI

Existential intelligence drives humans to question purpose, explore meaning, and evaluate the long-term impact of decisions. It is inherently reflective, enabling leaders to balance personal and organizational goals with a sense of broader societal contribution.

Philosophical AI uses logic to navigate abstract reasoning and ethical dilemmas. It provides analytical insights into complex questions but lacks emotional depth, cultural sensitivity, and the ability to introspect, limiting its capacity to align purpose with human values.

Existential Intelligence and Philosophical AI for Customer Growth

- **Tackle Ethical Dilemmas:** Use Philosophical AI to present logical frameworks for ethical decisions, while humans apply emotional and contextual insight to navigate complex moral challenges.
- **Explore Long-Term Impacts:** Pair Philosophical AI's analytical reasoning with human existential intelligence to evaluate the far-reaching implications of innovations and policies.
- **Support Reflective Leadership:** Use AI to provide logical frameworks for abstract concepts like customer evolution and societal shifts, enabling human leaders to refine their vision and align business goals with long-term customer growth.

Reflection

BEYOND THE MACHINE

As we approach a future increasingly shaped by advanced technology, the question is no longer what AI can do but what we, as humanity, choose to become alongside it. For all its computational brilliance and transformative potential, AI remains bound by its limitations—mimicry without meaning, knowledge without wisdom. Its strength lies in its precision, yet its power tempts us to relinquish our responsibility to understand and connect.

AI is no longer seen merely as a tool but as an infallible entity capable of salvation.

This belief system echoes religious fervor, seen in projects like a Swiss church's "AI Jesus," offering spiritual guidance through artificial means. Such endeavors risk blending spirituality with the mechanistic promises of AI, eroding the human-centered core of morality and faith. By placing AI on a pedestal of infallibility, we ignore its inherent flaws and biases, treating it as an unquestionable authority.

What if we resist the allure of convenience and speed? What if we refuse to let algorithms–guided by the seductive AI Ideology–define the edges of our imaginations, relationships, and sense of purpose? The machines we build may process endlessly, but they cannot ponder the stars, dream of better tomorrows, or find transcendence in the unpredictable beauty of existence.

H.G. Wells captured the fragility of human advancement in *The Time Machine*:

> "It is a law of nature we overlook, that intellectual versatility is the compensation for change, danger, and trouble. An animal perfectly in harmony with its environment is a perfect mechanism. Nature never appeals to intelligence until habit and instinct are useless. There is no intelligence where there is no change and no need of change."

Wells' vision of the distant future where humanity's reliance on technology would lead to stagnation, decay, and collapse is a poignant warning. What begins as convenience may end as complacency, and in that complacency lies the seed of downfall.

There is something sacred in the imperfections of human thought and feeling. It is in our mistakes that we grow, in our doubts that we seek, and in our vulnerabilities that we connect. AI, for all its sophistication, cannot replicate the ineffable—grief turned to gratitude, struggle transformed into meaning, or love shared across the chasm of difference.

As we integrate AI into our lives, we must remember that tools, no matter how advanced, are an extension of the hands that wield them. The paths AI illuminates are the ones we program, yet its brilliance dims in the absence of the human soul. Let us not measure progress in efficiency alone but in the depth of our compassion, the breadth of our imagination, and the courage to question the course we chart.

The future is neither dystopia nor utopia but actually a canvas upon which we paint. Will we allow the brush to slip into the grasp of the machine, or will we guide it with intention, blending the colors of logic and love, science and spirit, ambition and humility?

The answers lie not in AI's circuitry but rather in the existential questions that hum beneath the surface of our technological triumphs. What does it mean to be human? What do we owe to one another, the planet, and the unknown? How do we create a world where innovation amplifies connection rather than eroding it?

Let us not look to AI to save us but to remind us of the urgency of saving ourselves. Perhaps, in the end, the ultimate intelligence is not artificial at all but the timeless and unquantifiable capacity of humanity to seek, feel, and hope. Beyond the machine lies the future we choose to create—a future where the soul of humanity defines the purpose of its tools.

ENDNOTES

Introduction

(2022, February 7). *Science Gathers Knowledge Faster Than Society Gathers Wisdom.* Quote Investigator. Retrieved December 31, 2024, from https://quoteinvestigator.com/2022/02/07/science-wisdom/

(2022, August 28). *About Howard.* Howard Gardner. Retrieved September 14, 2024, from https://www.howardgardner.com/

(2023, June 4). *The Untapped Power of Consumer Insights.* Quid. Retrieved September 14, 2024, from https://www.quid.com/adweek-consumer-insights

(2024, December 25). *Salesforce TV Spot, "Rain" Featuring Matthew McConaughey, Woody Harrelson.* ISpot.tv. Retrieved December 31, 2024, from https://www.ispot.tv/ad/fPEh/salesforce-rain-featuring-matthew-mcconaughey-woody-harrelson

Andreessen, M. (2023, June 6). *Why AI Will Save the World.* Andreessen Horowitz. Retrieved December 31, 2024, from https://a16z.com/ai-will-save-the-world/

Cicek, M., Gursoy, D., & Lu, L. (2024). Adverse impacts of revealing the presence of "Artificial Intelligence (AI)" technology in product and service descriptions on purchase intentions: The mediating role of emotional trust and the moderating role of perceived risk. *Journal of Hospitality Marketing & Management, 34*(1), 1-23. https://doi.org/10.1080/19368623.2024.2368040

Deighton, K. (2024, November 25). *AI Ads Can Look Weird. Brands Like Coca-Cola Are Making Them Anyway.* The Wall Street Journal. Retrieved November 26, 2024, from https://www.wsj.com/articles/ai-ads-can-look-weird-brands-like-coca-cola-are-making-them-anyway-04331697

Di Placido, D. (2024, November 16). *Coca-Cola's AI-Generated Ad Controversy, Explained.* Forbes. Retrieved November 20, 2024, from https://www.forbes.com/sites/danidiplacido/2024/11/16/coca-colas-ai-generated-ad-controversy-explained/

Dot-com bubble. (2024, December 30). In *Wikipedia.* https://en.wikipedia.org/wiki/Dot-com_bubble

ELIZA. (2024, December 30). In *Wikipedia.* https://en.wikipedia.org/wiki/ELIZA

Gillett, K. (2019, April 10). *10 inspiring quotes from Lebanese poet Kahlil Gibran.* The National News. Retrieved October 20, 2024, from https://www.thenationalnews.com/arts/10-inspiring-quotes-from-lebanese-poet-kahlil-gibran-1.847136

Gray Widder, D., & Hicks, M. (2024, November 20). *Watching the Generative AI Hype Bubble Deflate.* Harvard. Retrieved December 15, 2024, from https://ash.harvard.edu/resources/watching-the-generative-ai-hype-bubble-deflate/

Jaffri, A., & Khandabattu, H. (2024, June 17). *Hype Cycle for Artificial Intelligence, 2024.* Gartner. Retrieved September 14, 2024, from https://www.gartner.com/en/documents/5505695

Maese, E. (2025, January 15). *Americans Use AI in Everyday Products Without Realizing It.* Gallup. Retrieved January 15, 2025, from https://news.gallup.com/poll/654905/americans-everyday-products-without-realizing.aspx

Marenus, M. (2024, February 2). *Howard Gardner's Theory of Multiple Intelligences.* Simply Psychology. Retrieved October 15, 2024, from https://www.simplypsychology.org/multiple-intelligences.html

Rosiek, T. (2024, December 26). *Why Predictive AI Is Essential to Federal Zero-Trust Security.* FedTech. Retrieved December 31, 2024, from https://fedtechmagazine.com/article/2024/12/why-predictive-ai-essential-federal-zero-trust-security

Toonkel, J. (2024, November 19). *New AI Entertainment Studio Gets Backing From Peter Chernin, Andreessen Horowitz.* The Wall Street Journal. Retrieved November 20, 2024, from https://www.wsj.com/business/media/new-ai-entertainment-studio-gets-backing-from-peter-chernin-andreessen-horowitz-8c840236

Tulfo, E. (2024, August 10). *Brands should avoid this popular term. It's turning off customers.* CNN. Retrieved September 1, 2024, from https://www.cnn.com/2024/08/10/business/brands-avoid-term-customers/index.html

Chapter 1

(2017, March 29). *Connecting Aspiration and Action: Nike Challenges the World to 'Just Do It'.* American Association of Advertising Agencies. Retrieved November 3, 2024, from https://www.aaaa.org/timeline-event/connecting-aspiration-action-nike-challenges-world-just/

(2018, April 19). *APA Dictionary of Psychology.* American Psychological Association. Retrieved November 3, 2024, from https://dictionary.apa.org/aspiration

(2024, November 27). *The impact of AI on content accuracy and reliability.* AIContentfy. Retrieved November 28, 2024, from https://aicontentfy.com/en/blog/impact-of-ai-on-content-accuracy-and-reliability

Astvansh, V. (2024, March 14). *Wendy's 'surge pricing' mess looks like a case study in stakeholder conflict.* The Conversation. Retrieved November 28, 2024, from https://theconversation.com/wendys-surge-pricing-mess-looks-like-a-case-study-in-stakeholder-conflict-225610

Cherry, K., MSEd (2024, May 6). *What Are Aspirations?* American Psychological Association. Retrieved November 3, 2024, from https://www.verywellmind.com/what-are-aspirations-5200942

Clark, T. (2024, January 3). *74 percent of U.S. Consumers are willing to share personal information with brands and retailers.* The Current. Retrieved November 11, 2024, from https://www.thecurrent.com/us-consumers-willing-share-personal-information-brands-retailers

Harrison Dupré, M. (22, September 18). *Experts: 90% of Online Content Will Be AI-Generated by 2026.* The Byte. Retrieved November 6, 2024, from https://futurism.com/the-byte/experts-90-online-content-ai-generated

Hart, C. S. (2016). *How Do Aspirations Matter?* Journal of Human Development and Capabilities, 17(3), 324–341. https://doi.org/10.1080/19452829.2016.1199540

Headley, C. (2023, October 4). *A Founder's Principle: The Power of Aspiration.* LinkedIn. Retrieved November 3, 2024, from https://www.linkedin.com/pulse/founders-principles-power-aspiration-courtney-headle

Henderson, R. M. (2018, February 12). *More and More CEOs Are Taking Their Social Responsibility Seriously.* Harvard Business Review. Retrieved November 29, 2024, from https://hbr.org/2018/02/more-and-more-ceos-are-taking-their-social-responsibility-seriously

Hood, C. (2023). *Customer Transformation.* CH Publishing. https://chrishood.com

Hutchinson, R., Novacek, G., Chin, V., & Falco, G. (2023, May 25). *Socially Transformative Business Is Smart Business.* BCG. Retrieved November 29, 2024, from https://www.bcg.com/publications/2023/socially-transformative-business-is-smart-business

Mitchell, K., Silverstein, A. E., & Sandoz, A. (2021, October 19). *Purpose—A beacon for growth.* Deloitte. Retrieved November 28, 2024, from https://www2.deloitte.com/us/en/insights/topics/marketing-and-sales-operations/global-marketing-trends/2022/brand-purpose-as-a-competitive-advantage.html

Petski, D. (2022, September 20). *UCLA Study Finds Teens Resoundingly Reject Aspirational Content On TV & Movies.* Deadline. Retrieved November 29, 2024, from https://deadline.com/2022/09/ucla-study-finds-teens-resoundingly-reject-aspirational-content-in-tv-or-movies-1235121262/

Saavendra, J., M.S. (2022, January 14). *Aspirations: Definition, Examples, & Insights.* Berkeley Well-Being Institute. Retrieved November 3, 2024, from https://www.berkeleywellbeing.com/aspirations.html

Tietjen, A. (2021, March 12). *Jane Lauder on Data and 'Aspirational Intelligence.'* Beauty Inc. Retrieved September 15, 2024, from https://wwd.com/feature/jane-lauder-estee-lauder-data-aspirational-intelligence-1234773181/

Wong, K. (2024, February 20). *Company core values: 25 inspiring examples.* Achievers. Retrieved November 29, 2024, from https://www.achievers.com/blog/company-core-value-examples/

Chapter 2

(2018, April 19). *APA Dictionary of Psychology: Aspiration.* American Psychological Association. Retrieved September 15, 2024, from https://dictionary.apa.org/aspiration

(2013, March 27). *Jurassic Park.* IMDb. Retrieved September 15, 2024, from https://www.imdb.com/title/tt0107290/

(2024, September 25). *E-commerce Market Size, Share, and Trends 2024 to 2034.* Precedence Research. Retrieved October 3, 2024, from https://www.precedenceresearch.com/e-commerce-market

(2024, October 23). *More Consumers are Willing to Share Data in Return for Personalization.* Porch Group Media. Retrieved October 30, 2024, from https://porchgroupmedia.com/blog/more-consumers-are-willing-to-share-data-in-return-for-personalization/

, S. (2024, July 25). *Yes, Your Car Is Spying On You. Here's What To Do About It.* U.S. News. Retrieved October 3, 2024, from https://www. usnews.com/opinion/articles/2024-07-25/yes-your-car-is-spying-on-you-heres-what-to-do-about-it

Goovaerts, D. (2024, February 5). *AI degeneration: Here's why aging models could be a problem.* Fierce Network. Retrieved November 7, 2024, from https://www.fierce-network.com/ai/ai-degeneration-heres-why-aging-models-could-be-problem

Haan, K., & Watts, R. (2023, July 20). Over 75% Of Consumers Are Concerned About Misinformation From Artificial Intelligence. Forbes. Retrieved September 15, 2024, from https://www.forbes. com/advisor/business/artificial-intelligence-consumer-sentiment/

Maake, K. (2023, April 27). *Numbers Game: Consumers are more willing to share their data if they get something in return.* Retail Brew. Retrieved October 3, 2024, from https://www.retailbrew.com/ stories/2023/04/27/numbers-game-consumers-are-more-willing-to-share-data-for-something-in-return

Milmo, D., & Hern, A. (2024, February 28). *Google chief admits 'biased' AI tool's photo diversity offended users.* The Guardian. Retrieved November 6, 2024, from https://www.theguardian.com/ technology/2024/feb/28/google-chief-ai-tools-photo-diversity-offended-users

Ng, A. (2024, August 13). *Texas sues General Motors over car data tracking.* Politico. Retrieved October 3, 2024, from https://www. politico.com/news/2024/08/13/texas-general-motors-car-data-tracking-00173877

Chapter 3

(2018, November 7). *Why B2B Leaders Should Get in Touch With Their Customers' Feelings.* Gallup. Retrieved November 30, 2024, from https://www.gallup.com/workplace/244607/why-b2b-leaders-touch-customers-feelings.aspx

(2022, July 26). *Most Companies Struggling to be Relevant to Their Customers, Accenture Report Finds*. Accenture. Retrieved October 15, 2024, from https://newsroom.accenture.com/news/2022/most-companies-struggling-to-be-relevant-to-their-customers-accenture-report-finds

(2024, November 14). *What are AI guardrails?* McKinsey & Company. Retrieved December 31, 2024, from https://www.mckinsey.com/featured-insights/mckinsey-explainers/what-are-ai-guardrails

Alvarez, G. (2024, October 21). *Gartner Top 10 Strategic Technology Trends for 2025*. Gartner. Retrieved November 23, 2024, from https://www.gartner.com/en/articles/top-technology-trends-2025

Bhatia, A. (2024, August 25). *When A.I.'s Output Is a Threat to A.I. Itself*. NY Times. Retrieved December 31, 2024, from https://www.nytimes.com/interactive/2024/08/26/upshot/ai-synthetic-data.html

Broustas, Y. (2024, December 11). *Agentic automation: The path to an orchestrated enterprise*. UI Path. Retrieved December 28, 2024, from https://www.uipath.com/blog/ai/agentic-automation-path-to-orchestrated-enterprise

Clark, T. (2024, January 3). *74 percent of U.S. consumers are willing to share personal information with brands and retailers*. The Current. Retrieved December 31, 2024, from https://www.thecurrent.com/us-consumers-willing-share-personal-information-brands-retailers

Dey, V. (2023, May 2). *92% of businesses use AI-driven personalization but consumer confidence is divided*. Venture Beat. Retrieved October 15, 2024, from https://venturebeat.com/ai/92-of-businesses-use-ai-driven-personalization-but-consumer-confidence-is-divided/

Faverio, M., & Tyson, A. (2023, November 23). *What the data says about Americans' views of artificial intelligence*. Pew Research Center. Retrieved October 15, 2024, from https://www.pewresearch.org/short-reads/2023/11/21/what-the-data-says-about-americans-views-of-artificial-intelligence/

Garrison, M. (2020, October 1). *What is Behavioral Intelligence?* Coeus Creative Group. Retrieved October 30, 2024, from https://www.coeuscreativegroup.com/2020/10/01/what-is-behavioral-intelligence/

Hood, C. (2023, August 20). *Customer Transformation.* ChrisHood.com. Retrieved November 23, 2024, from https://chrishood.com/customer-transformation

Kim, T., Lee, H., Kim, M. Y., Kim, S., & Duhachek, A. (2023). AI increases unethical consumer behavior due to reduced anticipatory guilt. *Journal of the Academy of Marketing Science, 51*, 785–801. https://doi.org/10.1007/s11747-021-00832-9

O'Donnell, J. (2024, November 26). *We need to start wrestling with the ethics of AI agents.* MIT Technology Review. Retrieved December 15, 2024, from https://www.technologyreview.com/2024/11/26/1107309/we-need-to-start-wrestling-with-the-ethics-of-ai-agents/

Oehler, P. (2022, October 26). *What is Behavioral Intelligence?* Retorio. Retrieved October 30, 2024, from https://www.retorio.com/blog/behavioral-intelligence

Taillard, M., & Giscoppa, H. (2013). *Psychology and Modern Warfare* (p. 183–187). Palgrave Macmillan New York. https://doi.org/10.1057/9781137347329_19

Wheeler, K. (2024, November 25). *Gartner: How Agentic AI is Shaping Business Decision-Making.* Technology Magazine. Retrieved December 1, 2024, from https://technologymagazine.com/articles/gartner-how-agentic-ai-is-shaping-business-decision-making

Witynski, M. (2021, August 11). *Behavioral economics, explained.* The University of Chicago. Retrieved October 15, 2024, from https://news.uchicago.edu/explainer/what-is-behavioral-economics

Chapter 4

(2023, May 24). *AI in the customer experience: The ups, downs, and up-and-coming opportunities.* SurveyMonkey. Retrieved October 20, 2024, from https://www.surveymonkey.com/resources/premium/cx-and-ai-report/

(2024, July 17). *How people really feel about the AI customer experience.* Spa Executive. Retrieved October 24, 2024, from https://spaexecutive.com/2024/07/17/how-people-really-feel-about-the-ai-customer-experience/

(2024, December 1). *Affectiva and Emotion AI.* Affectiva. Retrieved December 31, 2024, from https://www.affectiva.com/emotion-ai/

(2024, December 6). *Experience is everything. Get it right.* PwC. Retrieved December 20, 2024, from https://www.pwc.com/us/en/services/consulting/library/consumer-intelligence-series/future-of-customer-experience.html

(2024, December 31). *Starbucks.* Starbucks. Retrieved December 31, 2024, from https://starbucks.com

Azenkot, T. (2024, April 23). *AI overload: The good, the bad, and the ugly impact on customer experience.* Retail Customer Experience. Retrieved October 24, 2024, from https://www.retailcustomerexperience.com/blogs/ai-overload-the-good-the-bad-and-the-ugly-impact-on-customer-experience/

Blasi, W. (2024, November 9). *Starbucks's new CEO is making changes — Here's what customers and industry experts think of them.* CBS News. Retrieved November 9, 2024, from https://www.marketwatch.com/story/starbuckss-new-ceo-is-making-changes-heres-what-customers-and-industry-experts-think-of-them-3f06b41e

Bradberry, T. (2022, June 20). *Why Emotional Intelligence Can Save Your Life?* Talent Smart EQ. Retrieved November 22, 2024, from https://www.talentsmarteq.com/emotional-intelligence-can-boost-your-career-and-save-your-life/

Ed, D. (2023, June 5). *When AI Customer Service Goes Horribly Wrong, Humans to the Rescue.* Ask Doctor Ed. Retrieved October 24, 2024, from https://www.askdoctored.com/post/when-ai-customer-service-goes-horribly-wrong-humans-to-the-rescue

Hultgren, K. (2024, February 9). *Discover CMO Jennifer Murillo Shares Strategy Behind Pre-Game Super Bowl Spot.* Chief Marketer. Retrieved November 14, 2024, from https://chiefmarketer.com/discover-cmo-jennifer-murillo-shares-strategy-behind-pre-game-super-bowl-spot/

Lee, A. M. (2024, October 31). *Here's how Starbucks' new CEO plans to overhaul its struggling cafes.* CBS News. Retrieved October 31, 2024, from https://www.cbsnews.com/news/starbucks-menu-coffee/

Li, B., Liu, L., Mao, W., Qu, Y., & Chen, Y. (2022). Voice artificial intelligence service failure and customer complaint behavior: The mediation effect of customer emotion. *Electronic Commerce Research and Applications, 59.* https://doi.org/10.1016/j.elerap.2023.101261

Marr, B. (2021, July 13). *The Next Frontier Of Artificial Intelligence: Building Machines That Read Your Emotions.* Bernard Marr & Co. Retrieved December 31, 2024, from https://bernardmarr.com/the-next-frontier-of-artificial-intelligence-building-machines-that-read-your-emotions/

Somers, M. (2019, March 8). *Emotion AI, explained.* MIT Sloan. Retrieved December 31, 2024, from https://mitsloan.mit.edu/ideas-made-to-matter/emotion-ai-explained

Wahba, P. (2024, October 31). *Starbucks' new CEO just delivered what employees and investors have been craving after years of drama and missteps.* Yahoo Finance. Retrieved November 22, 2024, from https://finance.yahoo.com/news/starbucks-ceo-just-delivered-employees-195504824.html

Zaltman, G. (2003, January 13). *The Subconscious Mind of the Consumer (And How To Reach It).* Harvard Business School. Retrieved November 12, 2024, from https://www.library.hbs.edu/working-knowledge/the-subconscious-mind-of-the-consumer-and-how-to-reach-it

Zorfas, A., & Leemon, D. (2016, August 29). *An Emotional Connection Matters More than Customer Satisfaction.* Harvard Business Review. Retrieved November 22, 2024, from https://hbr.org/2016/08/an-emotional-connection-matters-more-than-customer-satisfaction

Chapter 5

(2016, June 7). *10 toys every kid wanted in 1976.* MeTV. Retrieved November 14, 2024, from https://www.metv.com/lists/10-toys-every-kid-wanted-in-1976

(2021, July 27). *What is computer vision?* IBM. Retrieved December 31, 2024, from https://www.ibm.com/think/topics/computer-vision

Cordasco, P. (2024, June 9). *Website personalization: Benefits & tips.* Salesforce. Retrieved November 23, 2024, from https://www.salesforce.com/marketing/personalization/guide/website/

Muir, J. (2016, May 29). *The Hottest Ride in Town: Remembering Marx's Green Machine.* Flashbak. Retrieved November 14, 2024, from https://flashbak.com/marxs-green-machine-60649/

Tatananni,, M. (2024, July 24). *AI SPY Virtual 'all-seeing' AI helpers that watch everything you do on computer screens revealed in warning over creepy upgrade.* The Sun. Retrieved December 31, 2024, from https://www.thesun.co.uk/tech/29449580/ai-assistants-microsoft-recall-data-privacy-concerns/

Chapter 6

(2024, February 22). *How iconic guitar maker Fender® increases customer engagement and lifetime value with its Fender Play® music lessons app.* Twilio. Retrieved December 31, 2024, from https://segment.com/customers/fender/

(2024, July 1). *Why Brands Need To Do More Than Just Sell!* Help Lama. Retrieved December 31, 2024, from https://helplama.com/brands-need-to-do-more-than-just-sell/

(2024, August 22). *50+ Food and Alcohol Delivery Statistics and Trends for 2024*. DoorDash. Retrieved December 22, 2024, from https://merchants.doordash.com/en-us/blog/food-delivery-statistics

(2024, December 2). *Learn the Guitar 3x Faster with Fender Play*. Fender. Retrieved December 31, 2024, from https://www.fender.com/play

(2024, December 10). *Online vs In-Store Shopping Statistics*. Capital One. Retrieved December 22, 2024, from https://capitaloneshopping.com/research/online-vs-in-store-shopping-statistics/?utm_source=chatgpt.com

(2024, December 28). *Why Do Companies Need To Be Proactive In Customer Service?* Help Lama. Retrieved December 31, 2024, from https://helplama.com/proactive-customer-service-survey/

Bayiates, A., Weiss, D., & Downey, E. (2023). *Deloitte's State of AI in the Enterprise* (4th ed.). Deloitte.

Bestsennyy, O., Gilbert, G., Harris, A., & Rost, J. (2021, July 9). *Telehealth: A quarter-trillion-dollar post-COVID-19 reality?* McKinsey & Company. Retrieved December 22, 2024, from https://www.mckinsey.com/industries/healthcare/our-insights/telehealth-a-quarter-trillion-dollar-post-covid-19-reality

Gothelf, J. (2023, March 13). *Case Study: How product management saved Fender guitars*. Jeff Gothelf. Retrieved December 31, 2024, from https://jeffgothelf.com/blog/how-product-management-saved-fender-guitars/

Tatananni,, M. (2024, July 24). *AI SPY Virtual 'all-seeing' AI helpers that watch everything you do on computer screens revealed in warning over creepy upgrade*. The Sun. Retrieved December 31, 2024, from https://www.thesun.co.uk/tech/29449580/ai-assistants-microsoft-recall-data-privacy-concerns/

US Department of Education (n.d.). *Objective 1: Increase Global and Cultural Competencies of All U.S. Students*. Retrieved December 28, 2024, from https://sites.ed.gov/international/objective-1-increase-global-and-cultural-competencies-of-all-u-s-students/

Chapter 7

(2019, October 1). *CGS Survey Reveals a Blend of Automation and Human Interaction Will Drive Meaningful Connections Between Customers and Brands.* CGS. Retrieved December 22, 2024, from https://www.cgsinc.com/en/news-events/cgs-2019-chatbots-survey-reveals-blend-automation-and-human-interaction-will-drive-meaningful-results

(2024, October 1). *AI in Social Media Market.* Markets and Markets. Retrieved December 22, 2024, from https://www.marketsandmarkets.com/Market-Reports/ai-in-social-media-market-92119289.html

(2024, October 20). *AI fueled by the neuroscience of music?* Lately. Retrieved October 20, 2024, from https://www.lately.ai/resources/about

Acar, O. A. (2023, June 6). *AI Prompt Engineering Isn't the Future.* Harvard Business Review. Retrieved October 20, 2024, from https://hbr.org/2023/06/ai-prompt-engineering-isnt-the-future

Al-Sibai, N. (2025, January 1). *People Are Disgusted by Facebook's Plan to Deploy AI-Powered "Users".* Futurism. Retrieved January 2, 2025, from https://futurism.com/disgusted-facebook-ai-users

Boyd, D. (2014, January 9). *A Creativity Lesson From Betty Crocker.* Psychology Today. Retrieved October 20, 2024, from https://www.psychologytoday.com/us/blog/inside-the-box/201401/creativity-lesson-betty-crocker

Coppola, D. (2023, December 13). *Share of customers preferring human connection vs. Automation in customer service in the United States in 2023, by generation.* Statista. Retrieved December 22, 2024, from https://www.statista.com/statistics/1428526/human-connection-vs-automation-in-customer-service-us/

Kane, C. (2024, July 9). *Hug it Out: The Power of the 8-Second Hug.* Dr. Christy Kane. Retrieved December 15, 2024, from https://www.drchristykane.com/post/hug-it-out-the-power-of-the-8-second-hug

Tenbarge, K. (2025, January 3). *Meta removes AI character accounts after users criticize them as 'creepy and unnecessary'*. NBC. Retrieved January 3, 2025, from https://www.nbcnews.com/tech/social-media/meta-ai-insta-shuts-character-instagram-fb-accounts-user-outcry-rcna186177

Chapter 8

(2019, November 5). *What's Your Customer Effort Score?* Gartner. Retrieved December 30, 2024, from https://www.gartner.com/smarterwithgartner/unveiling-the-new-and-improved-customer-effort-score

(2024, September 4). *The State of AI In Customer Success, 2024 Report*. Gainsight. Retrieved December 30, 2024, from https://www.gainsight.com/resource/the-state-of-ai-in-customer-success-2024-report/

(2024, December 6). *Experience is everything. Get it right*. PwC. Retrieved December 26, 2024, from https://www.pwc.com/us/en/services/consulting/library/consumer-intelligence-series/future-of-customer-experience.html

Cherniak, K. (2024, December 26). *AI in Customer Service Statistics: 50+ Actionable Insights for your Technology-Driven Business Strategy*. Customer Success Collective. Retrieved December 30, 2024, from https://masterofcode.com/blog/ai-in-customer-service-statistics

Dasgupta, M. (2024, July 9). *Top 17 Customer Success Metrics & How to Track Them*. Hiver. Retrieved December 14, 2024, from https://hiverhq.com/blog/customer-success-metrics

Fokina, M. (2024, October 31). *10+ Crucial AI Customer Service Statistics (2024)*. Tidio. Retrieved November 12, 2024, from https://www.tidio.com/blog/ai-customer-service-statistics/

Hood, C. (2023, October 5). *Unlocking Growth Through Understanding Customer Aspirations*. Chris Hood. Retrieved December 30, 2024, from https://chrishood.com/unlocking-growth-through-understanding-customer-aspirations/

Lakhwara, L. (2023, September 22). *AI and automation in customer success.* Customer Success Collective. Retrieved December 30, 2024, from https://www.customersuccesscollective.com/understanding-the-best-ways-to-use-artificial-intelligence-in-your-customer-success-strategies/

Lohia, N. (2024, December 23). *7 Different Ways to Measure Customer Satisfaction.* Hiver. Retrieved December 30, 2024, from https://hiverhq.com/blog/measuring-customer-satisfaction

Marr, B. (2024, August 15). *5 Common Generative AI Misconceptions.* Bernard Marr. Retrieved November 12, 2024, from https://bernardmarr.com/5-common-generative-ai-misconceptions/

Reichheld, F. F. (2003, December 1). *The One Number You Need to Grow.* Harvard Business Review. Retrieved December 22, 2024, from https://hbr.org/2003/12/the-one-number-you-need-to-grow

Chapter 9

(2022, June 15). *The End of AI Ambiguity.* InRule. Retrieved December 3, 2024, from https://inrule.com/resources/reports/report-the-end-of-ai-ambiguity/

(2024, January 30). *2024 retail industry outlook: Looking for loyalty in all the right places.* Deloitte. Retrieved December 3, 2024, from https://www2.deloitte.com/us/en/pages/consumer-business/articles/retail-distribution-industry-outlook.html

(2024, February 28). *#BrandsGetReal: What consumers want from brands in a divided society.* Sprout Social. Retrieved November 14, 2024, from https://sproutsocial.com/insights/data/social-media-connection/

(2024, October 30). *Majority of consumers don't trust brands when it comes to AI, according to Lippincott.* Lippincott. Retrieved December 30, 2024, from https://www.prnewswire.com/news-releases/majority-of-consumers-dont-trust-brands-when-it-comes-to-ai-according-to-lippincott-302290687.html

(2024, December 3). *Brand Loyalty Statistics*. Capital One. Retrieved December 3, 2024, from https://capitaloneshopping.com/research/brand-loyalty-statistics/

(2024, December 6). *Why invest in AI ethics and governance?* IBM. Retrieved December 7, 2024, from https://www.ibm.com/thought-leadership/institute-business-value/en-us/report/roi-ai-ethics

Boehm, J., Grennan, L., Singla, A., & Smaje, K. (2022, September 12). *Why digital trust truly matters*. McKinsey. Retrieved October 23, 2024, from https://www.mckinsey.com/capabilities/quantumblack/our-insights/why-digital-trust-truly-matters

Clark, S. (2022, July 19). *Do Your Customers Trust Your AI?* CMSWire. Retrieved December 3, 2024, from https://www.cmswire.com/customer-experience/do-your-customers-trust-your-ai/

Cordasco, P. (2024, June 9). *Website personalization: Benefits & tips*. Salesforce. Retrieved November 23, 2024, from https://www.salesforce.com/marketing/personalization/guide/website/

Delisi, R. (2024, March 11). *Why delighting customers doesn't pay*. Zendesk. Retrieved December 30, 2024, from https://www.zendesk.com/blog/delight-doesnt-pay/

Hironde, J. B. (2023, August 31). *AI's Impact On The Future Of Consumer Behavior And Expectations*. Forbes. Retrieved October 23, 2024, from https://www.forbes.com/councils/forbestechcouncil/2023/08/31/ais-impact-on-the-future-of-consumer-behavior-and-expectations/

McKendrick, J. (2024, December 27). *Why ethics is becoming AI's biggest challenge*. ZDNet. Retrieved December 28, 2024, from https://www.zdnet.com/article/why-ethics-is-becoming-ais-biggest-challenge/
Moral intelligence. (2024, March 26). In *Wikipedia*. https://en.wikipedia.org/wiki/Moral_intelligence

Overby, S. (2020, October 28). *The state of Artificial Intelligence (AI) ethics: 14 interesting statistics*. The Enterprisers Project. Retrieved December 28, 2024, from https://enterprisersproject.com/article/2020/10/artificial-intelligence-ai-ethics-14-statistics

Solomons, M. (2023, July 14). *100 branding statistics, global impact, and consumer perception*. Linearity. Retrieved December 3, 2024, from https://linearity.io/blog/branding-statistics/

Solomons, M. (2024, August 13). *2024 Consumer Loyalty Survey Insights to evolve brand loyalty programs*. Deloitte. Retrieved December 3, 2024, from https://www2.deloitte.com/us/en/pages/consulting/articles/brand-loyalty-program-consumer-behavior.html

Stone, K. (2024, May 15). *The State of Consumer Trust in the AI Era*. Get VOIP. Retrieved December 30, 2024, from https://getvoip.com/blog/consumer-trust-in-ai/

Wood, C. (2024, November 12). *Consumers are underwhelmed by AI experiences*. MarTech. Retrieved December 30, 2024, from https://martech.org/consumers-are-underwhelmed-by-ai-experiences/

Chapter 10

(2024, July 24). *Experian Named a 2024 Global Fintech Leader by Center for Financial Professionals for Fifth Consecutive Year*. Experian. Retrieved November 30, 2024, from https://www.experianplc.com/newsroom/press-releases/2024/experian-named-a-2024-global-fintech-leader-by-center-for-financ

(2024, October 21). *Workplace Skills Survey*. Deloitte. Retrieved December 22, 2024, from https://www2.deloitte.com/us/en/pages/about-deloitte/articles/human-skills-lacking-in-tech-driven-world.html

Anthony, S. D. (2024, October 24). *Tackling Disruption Playfully*. MIT Sloan. Retrieved December 22, 2024, from https://sloanreview.mit.edu/article/tackling-disruption-playfully/

Mankins, M. (2017, March 1). *Great Companies Obsess Over Productivity, Not Efficiency*. Harvard Business Review. Retrieved December 22, 2024, from https://hbr.org/2017/03/great-companies-obsess-over-productivity-not-efficiency

Ransbotham, S., Khodabandeh, S., Fehling, R., Lafountain, B., & Kiron, D. (2019, October 15). *Winning With AI*. MIT Sloan. Retrieved December 22, 2024, from https://sloanreview.mit.edu/projects/winning-with-ai/

Chapter 11

(2023, April 9). *Committed to being A Force for Growth and A Force for Good*. P&G. Retrieved November 30, 2024, from https://www.pg.co.uk/blogs/committed-to-being-a-force-for-growth-and-a-force-for-good/

(2024, January 20). *Collaboration, Emotional Intelligence, and Adaptability in Tech SMEs*. CyNam. Retrieved December 31, 2024, from https://cynam.org/resources/collaboration-emotional-intelligence-and-adaptability-in-tech-smes/

(2024, April 16). *2024 Best Brands for Social Impact*. Forbes. Retrieved December 1, 2024, from https://www.forbes.com/lists/best-brands-social-impact/

(2024, October 21). *AI Swarms: The Future of Teamwork?* PYMNTS. Retrieved December 31, 2024, from https://www.pymnts.com/artificial-intelligence-2/2024/ai-swarms-the-future-of-teamwork/

Bough, V., Breuer, R., Maechler, N., & Ungerman, K. (2020, October 27). *The three building blocks of successful customer-experience transformations*. McKinsey & Company. Retrieved November 6, 2024, from https://www.mckinsey.com/capabilities/growth-marketing-and-sales/our-insights/the-three-building-blocks-of-successful-customer-experience-transformations

Bouty, L. (2019, July 24). *Marketing Canvas - Aspirations*. Laurent Bouty. Retrieved December 1, 2024, from https://laurentbouty.com/blog/2019/marketing-canvas-aspirations

El Baroudi, S., Khapova, S. N., Fleisher, C., & Jansen, P. G. W. (2018). How Do Career Aspirations Benefit Organizations? The Mediating Roles of the Proactive and Relational Aspects of Contemporary Work. *Frontiers in Psychology*. https://doi.org/10.3389/fpsyg.2018.02150

Kaplan, J. (2925, January 7). *More Speech and Fewer Mistakes.* Meta. Retrieved January 8, 2025, from https://about.fb.com/news/2025/01/meta-more-speech-fewer-mistakes/

Kavanagh, D. (2019, March 29). *Aspirational Consumers: Why they Matter and What Brands Should Know.* GWI. Retrieved November 30, 2024, from https://blog.gwi.com/trends/why-aspirational-consumers-matter/

O'Brien, D., Main, A., Kounkel, S., & Stephan, A. R. (2019, October 15). *Purpose is everything.* Deloitte. Retrieved December 3, 2024, from https://www2.deloitte.com/us/en/insights/topics/marketing-and-sales-operations/global-marketing-trends/2020/purpose-driven-companies.html

Chapter 12
(2024, December 6). *Experience is everything. Get it right.* PwC. Retrieved December 8, 2024, from https://www.pwc.com/us/en/services/consulting/library/consumer-intelligence-series/future-of-customer-experience.html

(2024, December 31). *Disney.* The Walt Disney Company. Retrieved December 31, 2024, from https://disney.com

Bass, A. (2023, December 12). *Five Ways LEGO has Brought Customers into the Company – and how This Steers CX Strategies.* CX Today. Retrieved December 11, 2024, from https://www.cxtoday.com/loyalty-management/five-ways-lego-has-brought-customers-into-the-company-and-how-this-steers-cx-strategies/

Cazalis, V., & Barragan-Jason, G. (2022, December 14). *Humans and nature: The distance is growing.* Science Daily. Retrieved December 28, 2024, from https://www.sciencedaily.com/releases/2022/12/221214092431.htm

Cherry, K. (2024, March 14). *Naturalistic Intelligence and the Power of Environmental Awareness.* Very Well Mind. Retrieved December 28, 2024, from https://www.verywellmind.com/naturalistic-intelligence-8608334

Gregory, L. (2023, February 10). *Introduction to Naturalistic Intelligence.* Brightwheel. Retrieved October 22, 2024, from https://mybrightwheel.com/blog/naturalistic-intelligence

Hood, C. (2024, March 15). *The Integral Role of Non-Customer-Facing Teams in Customer-Centric Success.* LinkedIn. Retrieved December 11, 2024, from https://www.linkedin.com/pulse/integral-role-non-customer-facing-teams-success-chris-hood-ijknc/

Junger, S. (1997, January 1). *The Perfect Storm.* Sebastian Junger. Retrieved November 30, 2024, from https://www.sebastianjunger.com/the-perfect-storm

MacDonald, B. (2024, October 9). *Disneyland ticket price increase breaks through $200 barrier.* SiliconValley.com. Retrieved October 15, 2024, from https://www.siliconvalley.com/2024/10/09/disneyland-ticket-price-increase-breaks-through-200-barrier/

Chapter 13

(2022, November 7). *Employees Seek—and Struggle—to Find Meaning at Work.* Harvard Business. Retrieved October 30, 2024, from https://www.harvardbusiness.org/insight/employees-seek-and-struggle-to-find-meaning-at-work/

(2023, November 27). *The Power of Purpose in the Workplace.* Great Place to Work. Retrieved October 30, 2024, from https://www.greatplacetowork.com/resources/reports/the-power-of-purpose-in-the-workplace

(2024, September 1). *Klarna's Shocking Transformation: Fewer Humans, Record Profits.* Saltmarch Media. Retrieved December 24, 2024, from https://saltmarch.com/insight/klarnas-shocking-transformation-fewer-humans-record-profits

Bringsjord, S., & Govindarajulu, N. S. (2024, September 1). *Artificial Intelligence.* Stanford Encyclopedia of Philosophy. Retrieved December 31, 2024, from https://plato.stanford.edu/archives/fall2024/entries/artificial-intelligence

Carter, L. (2024, June 19). *The Connection Between Employee Satisfaction And Customer Satisfaction*. Best Practice Institute. Retrieved October 30, 2024, from https://blog.bestpracticeinstitute. org/employee-satisfaction-and-customer-satisfaction/

Dhingra, N., Samo, A., Schaninger, B., & Schrimper, M. (2021, April 5). *Help your employees find purpose—Or watch them leave*. McKinsey & Company. Retrieved October 30, 2024, from https:// www.mckinsey.com/capabilities/people-and-organizational-performance/our-insights/help-your-employees-find-purpose-or-watch-them-leave

Ellingrud, K., Sanghvi, S., Dandona, G. S., Madgavkar, A., Chui, M., White, O., & Hasebe, P. (2023, July 26). *Generative AI and the future of work in America*. McKinsey & Company. Retrieved December 24, 2024, from https://www.mckinsey.com/mgi/our-research/generative-ai-and-the-future-of-work-in-america

Fioretti, L., La Croce, C., Siviero, A., & Clemmons, E. (2024, October 21). *The AI Pivot Towards Becoming an AI-Fueled Business*. IDC. Retrieved December 24, 2024, from https://www.idc.com/getdoc.jsp?containerId=prAP52668124

Noichl, M. (2024, October 28). *The purpose of philosophical AI will be: To orient ourselves in thinking*. AI Impacts. Retrieved December 31, 2024, from https://aiimpacts.org/the-purpose-of-philosophical-ai-will-be-to-orient-ourselves-in-thinking/

Ostberg, R. (2024, November 11). *Transhumanism*. Encyclopedia Britannica. Retrieved December 31, 2024, from https://www. britannica.com/topic/transhumanism

Samuel, S. (2019, November 15). *How biohackers are trying to upgrade their brains, their bodies — And human nature*. Vox. Retrieved December 31, 2024, from https://www.vox.com/future-perfect/2019/6/25/18682583/biohacking-transhumanism-human-augmentation-genetic-engineering-crispr

Samuel, S. (2024, November 21). *Shannon Vallor says AI does present an existential risk — But not the one you think*. Vox. Retrieved December 31, 2024, from https://www.vox.com/future-perfect/384517/shannon-vallor-data-ai-philosophy-ethics-technology-edinburgh-future-perfect-50

Shukla, M. (2024, December 23). *What companies do now will determine their future in the Intelligent Age*. World Economic Forum. Retrieved December 24, 2024, from https://www.weforum.org/stories/2024/12/companies-determine-future-in-intelligent-age/

INDEX

www.ingramcontent.com/pod-product-compliance
Lightning Source LLC
Chambersburg PA
CBHW031501180326
41458CB00044B/6658/J